FOOD SAFETY AND QUALITY

TRADE CONSIDERATIONS

ORGANISATION FOR ECONOMIC CO-OPERATION AND DEVELOPMENT

ORGANISATION FOR ECONOMIC CO-OPERATION AND DEVELOPMENT

Pursuant to Article 1 of the Convention signed in Paris on 14th December 1960, and which came into force on 30th September 1961, the Organisation for Economic Co-operation and Development (OECD) shall promote policies designed:

- to achieve the highest sustainable economic growth and employment and a rising standard of living in Member countries, while maintaining financial stability, and thus to contribute to the development of the world economy;
- to contribute to sound economic expansion in Member as well as non-member countries in the process of economic development; and
- to contribute to the expansion of world trade on a multilateral, non-discriminatory basis in accordance with international obligations.

The original Member countries of the OECD are Austria, Belgium, Canada, Denmark, France, Germany, Greece, Iceland, Ireland, Italy, Luxembourg, the Netherlands, Norway, Portugal, Spain, Sweden, Switzerland, Turkey, the United Kingdom and the United States. The following countries became Members subsequently through accession at the dates indicated hereafter: Japan (28th April 1964), Finland (28th January 1969), Australia (7th June 1971), New Zealand (29th May 1973), Mexico (18th May 1994), the Czech Republic (21st December 1995), Hungary (7th May 1996), Poland (22nd November 1996) and Korea (12th December 1996). The Commission of the European Communities takes part in the work of the OECD (Article 13 of the OECD Convention).

Publié en français sous le titre :

QUALITÉ ET SÉCURITÉ ALIMENTAIRES :
LES DIMENSIONS COMMERCIALES

FOREWORD

This study examines international trade conflicts arising from food safety and quality issues from an economic rather than legal perspective. The study summarises the key international agreements, illustrates the range and nature of current disputes, reviews the potential contributions of economic analysis to conflict resolution, and identifies areas requiring further analysis. As such, it provides useful background to governments, industry and academics interested in food safety, regulation and trade.

The study complements two earlier OECD Secretariat papers examining current issues related to food safety and quality -- *Costs and Benefits of Food Safety Regulations: Fresh Meat Hygiene Standards in the United Kingdom* [OECD/GD(97)149] and *Uses of Food Labelling Regulations* [OECD/GD(97)150]. Both these documents are available for free at www.oecd.org/agr/publications/index1.htm.

This study was originally prepared by the consultants Jean-Christophe Bureau, Estelle Gozlan and Stéphane Marette of the Institut national de la recherche agronomique (France) and subsequently revised by Wayne Jones of the Food, Agriculture and Fisheries Directorate, following detailed review by OECD countries. It is published as a consultant's report on the responsibility of the Secretary-General of the OECD as recommended by the Committee for Agriculture and the Trade Committee.

TABLE OF CONTENTS

Boxes

Food Safety and Quality: Trade Considerations

Executive Summary

This report examines the trade conflicts arising from food safety and quality issues. It summarises the key international agreements, illustrates the range and nature of current disputes, reviews the potential contributions of economic analysis to conflict resolution, and identifies areas requiring further analysis. The report does not make policy recommendations.

Main observations from the report:

- ***On assessment of risk***: The Sanitary and Phytosanitary (SPS) Agreement explicitly requires Members to base their SPS measures on risk assessment as appropriate to the circumstances of the risk to human, animal or plant health. However, there is no agreement on what constitutes justifiable risk or "acceptable risk" as mentioned in the SPS Agreement. Nor is there any agreement on the importance to be given to risk analysis, or on what is meant by the term "risk", or on methodology.

- ***On recognition of quality attributes:*** There is considerable disagreement on quality attributes, such as the nutritional content and production methods, and on the extent to which they may legitimately be the subject of international regulation. These notions of product quality are ill matched to the more restrictive approach adopted *de facto* internationally. The stance of the SPS Agreement is to take into consideration only sanitary and phytosanitary factors.

- ***On environmental measures:*** A question mark remains as to the practical scope of Article XX of GATT, which mentions the protection of resources as potentially acceptable grounds for commercial discrimination, and of the Technical Barriers to Trade (TBT) Agreement, which mentions it as possible grounds for exceptions to international standards. There are also a number of other regional and multilateral environmental agreements, and many ambiguities remain concerning the respective roles of United Nations agreements, regional agreements and the WTO.

- ***On cultural differences:*** A country may introduce regulations that are more stringent than international standards on cultural grounds only under very limited conditions. Ignoring consumer values may result in a falling away of their support for the process of trade liberalisation, however, giving consideration to ethical, cultural or moral arguments could be used to cover a potentially unlimited number of exceptions to trade.

- ***On Intellectual property rights:*** Intellectual property rights to biological resources remain a grey area. There is as yet no multilateral framework within which emerging controversies over patents on living creatures can be discussed, points of view harmonised and clear rules drawn up.

- ***On the precautionary principle:*** Provisions of the SPS Agreement regarding precaution, are much more restrictive than what some consumer groups often mean when they invoke the "precautionary principle", suggesting that there may be a fundamental ambiguity between the expectations of certain groups in society and practical measures.

- ***On Codex Alimentarius:*** The standards accepted by scientists do not always have an indisputable scientific foundation: some have had to be completely revised at various times, and scientific "unanimity" is seldom achievable. International standards are now put to the vote at the *Codex*, and some are passed by a small majority. Not all countries are willing to acknowledge the legitimacy of risk thresholds imposed on them in this way.

- ***On the contributions of economic analysis:*** An assessment of sanitary risk alone, provides no basis whatsoever for deducing the cost of that risk to society. The SPS Agreement distinguishes between public health aspects and animal and plant health. Only for the latter is it foreseen that "relevant economic factors" should be taken into account.

Suggested considerations for further analysis:

- The interpretation of the precautionary principle, in particular as set out in Article 5.7 of the SPS Agreement, on the basis of which Members may adopt sanitary and phytosanitary measures on a provisional basis while seeking additional information, where scientific evidence is insufficient.

- A review of different national approaches to emerging social concerns and consumer preferences that would focus on how food safety and quality regulations are being implemented (e.g. a comparative analysis of regulatory mechanisms governing agricultural biotechnology in Member countries).

- Examine the compliance costs to industry, which can have a significant impact on the viability and international competitiveness of firms (especially small and medium size operations) and can be passed on to consumers in the form of higher food costs.

- The growing incidence of private standards justifies an examination of the changing role of private industry in food safety and quality, including the potential trade implications (e.g. market access, harmonisation, transparency, competition).

FOOD SAFETY AND QUALITY ISSUES:
TRADE CONSIDERATIONS

1.   Introduction

The emergence of food safety and quality as a trade issue relates to a background of profound changes. The OECD population has become predominantly urban, and because of modern consumption habits, such as catering, ready to cook and convenience food, consumers are more dependant on public authorities for food safety than in traditional rural societies. Technological progress and processing innovations, economic growth and increased international trade have put on the shelves many new products, requiring a better mastering of food safety in the food chain than traditional methods of conservation. Moreover, as incomes rise, consumers become more demanding and are more prepared to pay for a regulatory regime that provides higher standards and minimises risks. Demands for increased regulation and stricter enforcement of existing regulations have gained momentum across OECD countries in recent years due to a number of highly publicised outbreaks of food-borne diseases (e.g. BSE, *E. Coli*, *Salmonella*, *Campylobateria* or *Listeria*). Biological hazard is specific because diseases, micro-organisms or parasites can spread very rapidly and develop into wide or epidemic contamination. Such characteristics imply that outbreaks may have high costs. Therefore, safety and sanitary considerations require the possibility for quick identification and isolation of an element of the chain when at risk. All these elements make food safety a more complex issue.

In fact, consumer concerns go well beyond basic food safety. The quality of food and how it is produced, animal welfare, the use of genetically modified organisms (GMOs), hormones and other growth promoters, cultural preferences, resource sustainability and protection of the environment have all become issues in the public debate over regulation of the food industry. New production methods driven by technology have added to consumer unease, fuelled by a growing mistrust in the use of some techniques (e.g. irradiation) and differences in the interpretation of science in terms of food regulation. It has become essential for governments to consider all potential risks to the safety and wholesomeness of food, at all stages of the food chain.

In general, policy reform is contributing to a gradual deregulation of the agricultural sector in some countries, but where food safety and quality are concerned increased regulation is the norm (OECD, 1997a). For many emerging public concerns, agro-food is one of the most visible sectors and pressures for government intervention are high. For example, Canada has recently established a new food agency with a broad mandate for health, safety and inspection responsibilities while a similar agency is in the process of being set up in France and another proposed in the United Kingdom. The United States has announced a new initiative to address the health risks of food consumption involving several federal agencies with related responsibilities and the authority of the USDA in this area has been enhanced. In the European Union, the Commission has issued a green paper aimed at triggering a wide debate on food policy and food law. In these initiatives, government regulation is not the only approach deserving consideration, with measures ranging from voluntary practice, codes of good conduct, private standards, labelling and economic incentives. Still, the issues are complex and the required policy response unclear. The appropriate response is especially difficult to ascertain in cases where there are strong consumer concerns and insufficient or uncertain scientific evidence of health risk.

There is a question about the impact of stricter food safety and quality measures on international trade; a need to distinguish between legitimate food safety measures and protectionism. Standards and procedures can facilitate trade, but they may also reduce international competition, distort trade and prevent firms, notably foreign firms, from entering the market if they are discriminatory. The form and level of regulations differ considerably with inconsistencies in risk assessment over time, among sectors and across countries. Although they are often viewed from a domestic perspective, regulations have significant trans-boundary implications.

As traditional barriers to trade come down, regulatory reform has become a more important trade issue (OECD, 1998). The implementation of the Uruguay Round, in particular the Sanitary and Phytosanitary (SPS) and Technical Barriers to Trade (TBT) Agreements, have provided significant momentum towards the use of international standards. Increased notification requirements for both TBT and SPS regulations and the desire to avoid WTO dispute settlement procedures should make countries more aware of and careful about the international impacts of their regulations, including those aimed at protecting human, animal or plant health.[1] A number of disputes involving

---

1. The major agricultural importing and exporting countries are conscientiously observing the transparency obligations although more than half the WTO members (mostly low or middle income countries) have not yet notified a

several OECD countries have already been brought before the WTO since the dispute settlement procedure was established in 1995. With the strengthening of international rules, increased trade in consumer food products and the growing use of biotechnology, trade conflicts over food regulatory issues and their reform could become more common. The economic stakes are high and such disputes are likely to remain a priority in the future trade agenda.

This report examines the trade conflicts arising from food safety and quality issues. The paper is divided into two main parts. The first part provides some illustrations of the range and nature of current disputes and discusses the underlying problems and policy challenges for the future (Sections 2 and 3). The second part of the paper reviews the economic literature in an attempt to discern how economic analysis can contribute to the resolution of these issues (Section 4) and identifies areas requiring further analysis (Section 5). An assessment of the trade implications of food safety and quality regulations and related policy recommendations are outside the scope of this "issues" paper, which aims at raising questions and underlining the challenges.

2. Regulations and trade

2.1. Definitions and concepts: product quality and safety

In the remainder of this report, the term "safety" refers to the risk of germs, toxins and pathogenic chemical residues to human health and of the spread of diseases or parasites that might affect plant or animal health. Here, the meaning of the term "quality" is broader in scope, covering all the attributes included in the consumer's utility function, *i.e.* all those factors which are likely to cause a consumer to prefer one good to another. Quality refers not only to safety-related and nutritional aspects but also to wholesomeness and product characteristics such as the taste, integrity and even authenticity of products. The concept of quality adopted here is therefore very general. It also includes ethical factors, whereby the consumer's satisfaction or dissatisfaction is not with the characteristics of the product but with the way in which it has been produced (Mahé, 1997). For example, a good produced in an environmentally harmful way or using child labour or in breach of the elementary rules of animal welfare may be of "low quality" for some consumers because it offends their personal values. This extensive definition of quality also includes individual

---

single measure. A total of 724 SPS measures have been notified by 52 WTO Member countries during the first two years that the Agreement has been in force.

sensibility to particular technologies, such as biotechnology or irradiation, and procedures (kosher or halal products), for cultural or religious reasons.

2.2. *International effects of regulation*

Government intervention is often required to protect consumers by ensuring that food products are irreproachable in safety terms and that consumers get the quality they pay for. Food quality and safety regulations are not only necessary for public health reasons, they are also a precondition for the trust of economic agents and the smooth working of markets.

But food quality and safety regulations also have an effect on trade (Henson, 1998). To begin with, regulations can affect production costs and hence a country's competitiveness on the world market. This is the case with regulations concerning production processes, such as compliance with certain environmental constraints or technical specifications.

Confidence that products are both safe and of high quality can favour trade by alleviating consumer fears about imported products and by ensuring fair competition with regard to quality. Sometimes, however, national regulations can pose problems for exporters. This is the case, for example, when various options exist for ensuring a similar level of consumer protection. In order to ensure that a product is safe, a government may consider banning certain techniques (*i.e.* controlling processes) or laying down maximum tolerance levels for residual pathogens. In biological terms, one option cannot be said to be systematically preferable.[2] But if one country's standards are based on the first option and another country's are based on the second, firms cannot always produce for both markets, since their exports come up against technical barriers and additional control costs (Roberts and DeRemer, 1997).

Differing tastes, incomes, willingness to pay for quality and acceptable levels of risk may also lead to differing regulations. Developing countries cannot always afford the same quality standards as developed countries, with the result that their firms may come up against regulations which constitute a *de facto* barrier to exports. (Often this leads to a two-tier system with exporting firms operating with higher standards than those producing for

---

2.   It has been shown in some cases that controlling processes could give negligible levels of bacteriological risk and an overall result at least equivalent to sterilisation of the end product (e.g. mineral waters bottled at source in comparison with sterilised waters; or Hazard Analysis Critical Control Point — HACCP —applied to meat, in comparison with irradiation).

the domestic market.) This situation also exists between OECD countries, where differing preferences for quality attributes are reflected in dissimilar regulations. In economic terms, it is possible to determine an optimum quality standard for each country, reflecting in particular a trade-off between cost and demand for quality (Antle, 1995; Viscusi *et al.*, 1995). It depends on the distribution of consumers' willingness to pay, and there is no reason why such an optimum standard should be the same in all countries. But different standards, albeit entirely justified in economic, social and political terms, can hamper the further development of international trade, in particular trade of bulk commodities.

Lastly, the possibility of technical or sanitary regulations being used to reduce competition from imported goods and protect local producers cannot be ignored. Numerous studies have described where special interest groups have argued for health and safety barriers to imports, despite evidence that risk levels were minimal (Hillman, 1978, 1997; Kramer, 1989; Roberts and Orden, 1997). Economists typically attribute the origin of such measures to rent-seeking by special interest groups (Stigler, 1975). In the case of environmental and safety standards, for example, political and economic pressures may influence thresholds and framing regulations (Magat *et al.*, 1986; Tullock, 1997; Powell, 1997).

There are few quantitative estimates of the impact of national regulations on trade or welfare, though government agencies have attempted to estimate the impact of third country regulations on the value of exports. Such studies generally cite specific cases where third country regulations make it difficult to market national products (Roberts and DeRemer, 1997; Thornsbury *et al.*, 1997, Commission, 1997; USITC, 1997; Johnson, 1997; Ndayisenga and Kinsey, 1994). Even though few precise estimates exist, and even though the figures that the various countries put forward are always arguable, overall these studies suggest that national product quality regulations can have a significant effect on agro-food trade.

2.3.   *Legal framework and international agreements*

As a result of multilateral negotiations, rules have been introduced aimed at minimising the negative effects of sanitary and phytosanitary and technical regulations on international trade. GATT panel decisions have established the general principle that international rules do not permit WTO members to restrict the imports of products on the basis of how they are produced (Vogel, 1995). The Sanitary and Phytosanitary Agreement (SPS), concluded in the context of the Uruguay Round, asserts the right of signatory countries to take the measures they deem necessary to protect human, animal or plant life or health, provided that such measures are based on scientific principles, are not maintained without sufficient scientific evidence, and are not applied in an arbitrary or unjustifiable way. The Agreement states that sanitary measures may not be used for protectionist purposes. National regulations shall be based on international standards or, if not, on scientific justification. The SPS Agreement encourages the harmonisation of SPS measures based on internationally accepted standards, guidelines or recommendations. It also encourages the signatories to conclude bilateral and multilateral agreements on the recognition of equivalent sanitary and phytosanitary measures of other WTO members (Annex 1).

The scope of the 1979 Technical Barriers to Trade Agreement (TBT) was also extended during the Uruguay Round. All WTO members have had to comply with it since the Marrakech Agreement, and a country may not reject panel conclusions simply because they are unfavourable. The TBT Agreement is wide-ranging and covers all technical regulations and standards except those falling under the SPS Agreement, including those relating to packaging and labelling (Annex 1). The Uruguay Round also resulted in the introduction of a notification procedure which acts as an early warning system when national TBT or SPS regulations would be liable to restrict trade. If the disagreement persists, the WTO may, at the request of the parties involved, set up a panel to examine the problem.

The standards drawn up by international bodies now serve as a benchmark. The SPS Agreement, for example, recommends that all WTO members base their SPS measures on the international standards, guidelines and recommendations developed by the relevant international organisations and their subsidiary bodies, in particular *Codex Alimentarius* or *Codex* (Annex 2), the International Plant Protection Convention or IPPC (Annex 3) and the International Office of Epizootics or OIE (Annex 4). Members' measures that are based on international standards are deemed to be in accordance with the SPS Agreement.[3] The SPS Agreement allows for higher standards, provided certain conditions are met. Members may introduce or maintain SPS measures that result in a higher level of protection than that achieved by the relevant international standards, if there is scientific justification, or if it is a consequence of a level of SPS protection deemed appropriate by the Member, based on an appropriate assessment of risks.

---

3. When a regulation complies with an international standard, there is no need to notify the WTO or to justify it against a challenge from another State.

2.4.   *Private standards*

The growth of private standards on top of public regulation effectively increases the range of food safety and quality standards with which suppliers -- domestic and foreign -- may have to cope with when entering specific markets. Voluntary approaches by the industry to manage food safety and quality issues have, in some cases, been adopted by governments and made mandatory (e.g. a number of OECD countries have required the food processing industry to apply Hazard Analysis Critical Control Point -- HACCP). The implementation of the HACCP, or similar regulatory programmes, can be trade facilitating if developed in a harmonised manner world-wide, but may hamper trade if introduced without consideration of other nations' adoption of the regime. This is of particular concern for developing countries -- where existing infrastructure may not allow for the adjustment needed to meet new requirements. At a minimum, developing countries may need additional time and assistance to put such systems in place.

Private food safety standards can take two main forms -- self-regulation and market regulation. In most cases such forms of regulation are not substitutes for government action, but co-exist with public regulation (OECD, 1997b). The key characteristic of private standards is that the primary responsibility for formulation and compliance of the standards rests with a private agency, for example a trade association or dominant player in the market, rather than a government agency. However, this does not imply that firms operating in the market have greater freedom to choose whether to comply with the standards than is the case with public regulation. The pressure on firms to comply due to market forces can be just as great as the threat of legal action by government enforcement agencies. In some countries, like New Zealand, these industry standards are enforced at government level. They have the force of law and, if complied with, carry government assurances, or attract penalties for non-compliance.

Self-regulation. Internationally, the most widely applied form of self-regulation is ISO 9000 certification. The International Standards Organisation (ISO) based in Geneva, develops standards which represent voluntary principles of good practice and the ISO 9000 series of standards detail internationally accepted procedures and guidelines to maintain a consistent quality in product design, production, installation and servicing, and practices for certification. ISO certification then involves a third party certifying that these aspects of a firm's quality management system is in accordance with the principles laid down by the standard. These standards are not intended to replace product safety or other regulatory requirements, but specify those elements that quality management systems must have to produce final products that consistently meet the required specification. The focus here is on system quality rather than the quality of the end product.

However, most forms of private food quality standards do not conform to a recognised national or international standard, but gain their credibility from acceptance by the industry in which they operate, in particular the key purchasers of the product. Quality schemes are an example of this type of private standard (e.g. British Meat Manufacturers Association members must comply with a comprehensive set of hygiene and manufacturing practice standards that extend well beyond the scope of public safety regulations). Food safety and hygiene standards can also be administered by industry-level organisations which are instituted as companies in their own right.   Such schemes which encompass all aspects of production from input supply through to final delivery in retail outlets, are increasingly setting industry standards.

Market regulation.   The distinction between self-regulation and market regulation is that whilst the standards laid down by the former are voluntary, the standards laid down by the latter are imposed on firms by the market in which they operate.   The most common form of market food standards are the specifications imposed on firms by their major customers (e.g. large retailers may impose strict product and process standards which exceed public regulations on suppliers of their own-label products). Accreditation to ensure the standard has been satisfied is regarded by the retailers as an entry requirement which must be satisfied before a transaction can take place. The standards laid down by individual retailers can of course differ and therefore suppliers may simultaneously have to be accredited against a number of different standards, some of which may conflict with one another. In each case the cost of the accreditation is borne by the supplier. This applies to both domestic suppliers and suppliers in other countries.

An interesting trend in recent years has been for public regulation to effectively follow the *de facto* standards laid down by private regulation. A prime example is the requirement for Hazard Analysis Critical Control Point (HACCP) control systems which is becoming a common feature of public food safety regulations.[4] HACCP was originally developed as a voluntary process-oriented management tool to help attain a performance goal of safe food. An important characteristic of this approach is its inherent flexibility; a number of general managerial principles are laid down which can then be adapted to suit the situation in which the system is being applied.

---

4.   For example, EU Directive 93/43 on general food hygiene requires the application of food control systems based on HACCP principles.

3. Trade conflicts

3.1. International disputes

International agreements have not solved all the problems that food quality (in the broad sense), and in particular food safety, raise with regard to international trade. The following examples are illustrative of current disputes relating to regulation.

Sanitary and veterinary controls. It is not easy to harmonise veterinary and sanitary regulations, especially between countries at different levels of development. Countries like India, for example, have voiced their displeasure at the sanitary measures implemented by some more developed countries. They regard sanitary regulations which block their exports to North America and Europe as unfair. Conversely, consumer organisations, fearing that imports from less developed countries will drag international sanitary and phytosanitary standards down, are expressing concerns about Article 10 of the SPS Agreement, which recommends taking account of the special needs of developing countries in the definition of standards (Silverglade, 1998). The experience of European Union Member countries suggests that mutual recognition of national standards is the only practical way to reach an agreement, since full harmonisation is often too long and costly (Vogel, 1995). However, regarding other national regulations as equivalent also raises difficulties, even between OECD countries. For example, the European Union and the United States have made progress toward an agreement, but the two sides have not yet been able to agree on mutual recognition of their poultry slaughter standards.

Bovine Spongiform Encephalopathy (BSE). Following the discovery in 1996 of a possible link between BSE and Creuzfeldt-Jacob disease in humans, restrictions were imposed on the import of beef and derived products. In 1997, the United States took measures against imports of "specific risk" products in recognition of the precautionary principle. In 1998, the European Commission proposed to ban the use of specified beef, sheep and goat offal (although not approved by the EU Council). In both cases, third countries protested against these precautionary measures because they restricted imports, including from regions hitherto untouched by the disease. This example illustrates how difficult it can be for international agreements to accommodate protective measures against potential risks or risks about which little is known. (The terms relating to the precautionary principle, as laid down in Article 5.7. of the SPS Agreement are, in this respect, sometimes interpreted differently by the various Member countries). This example also raises the question of the practical limitations of the general GATT principle that standards for imports and domestic products must be similar, when a disease is only present in one country, or group of countries. However, there is some scope to address such problems under Article 6 of the SPS Agreement dealing with regionalisation.

Chemical residues. National regulations on authorised pesticide residues differ widely. However, the fact that it is difficult to measure the risks in this area makes any attempt to define standards highly controversial (Mazurek, 1996). Even national regulations applied even-handedly to domestic and imported products can have an effect on trade, especially if the chemical substances are not used in the country concerned. This is the case with procymidone, for example, a fungicide which is the subject of controversy in the wine-making industry. As the fungus against which procymidone is effective does not pose a problem in Californian vineyards there is no reason to use procymidone there. But low tolerance levels for residues [in the absence of an application of maximum residual levels (MRL)] would indirectly limit imports of wine from other countries which need to use the product because of their climate.[5] This highlights the possible trade effects of national regulations, even when they are applied to imported products in a non-discriminatory way.

Regulations on hormone use. The dispute between the European Union and the United States and Canada on the use of growth hormones in cattle has been going on for ten years. The EU's prohibition of the use of such substances has the effect of limiting imports from third countries where they are used. The dispute was settled on appeal in 1998. The Appellate Body concluded that the measure at dispute should have been based on an assessment as appropriate to the circumstances of the risks to human health and that, in that respect, the scientific evidence presented by the European Community was not sufficiently specific. BST is another hormone used in a number of countries, including the United States, to increase milk yields. In 1990, the EU imposed a moratorium on the use of BST until the end of 1999 (though without banning imports of dairy products from countries where BST is allowed). Canada has also not approved the use of BST in milk production. In accordance with its "human health" stance, the *Codex Alimentarius* approached the problem of BST and growth hormones from the standpoint not of farming practice but of measurable residues, which proved to be low in both cases. It was then argued that economic and social factors and consumer reluctance should be taken into

---

5.   The "Delaney clause" in US legislation, revised in 1996, banned products if a residue of a product susceptible of causing cancer was identifiable, even though technological progress meant that the detection threshold had fallen to the level of a single molecule, legitimising the ban of a good even though the risk associated with its consumption was close to zero (see Antle, 1995).

account. But the decision made at the *Codex* to defer and reconsider the BST case, was not on account of these arguments.[6] The differences within the *Codex* on BST raise the issue of the consideration that should be given to factors other than scientific evidence when sanitary regulations are being defined.

HACCP specifications. Hazard Analysis Critical Control Point (HACCP) is a set of principles which make it possible to identify hazards and determine the points in the transformation process at which it is most useful to carry out controls in order to eliminate the risk of contamination with pathogens. It establishes target and acceptable hazard levels and defines the corresponding tests. The broad principles of the method have been specified by the *Codex Alimentarius*, and many countries now use the HACCP method, originally voluntary but in fact increasingly compulsory. However, divergencies between the methods required within the HACCP system can pose problems at an international level. Within the broad principles, some countries have drawn up more stringent specifications than others, with highly detailed and complex plans (e.g. for meat in the United States). This creates tension, because countries do not always recognise each other's HACCP systems (Caswell and Hooker, 1996). International trade could be restricted if the countries with the strictest systems require equivalent inspection systems in other countries. This highlights the possible effects of differing specifications, even between countries which share broadly equivalent food safety concerns (Doussin, 1998).

Industry standards. The role that should be left to private operators in devising workable standards is a source of disagreements within and between countries. Producers want to be given greater freedom in the way they produce high quality food and point to the costs that highly specific regulations impose on the production process. They find it hard to understand why consumers and public authorities interfere so much in the definition of standards which in other

---

6. The consumer representative and several countries argued that consumers were opposed to the use of BST and that BST improved neither the quality nor the health characteristics of milk, and asked to be allowed to ban it. The European Union asked for "legitimate factors other than scientific analysis" to be taken into consideration. But the vote to defer the decision was taken because some Delegations had contributed scientific evidence which raised questions about the weakening of the immune systems of animals treated with BST. They argued that this could increase the risk of infection and hence the need for treatment and hence levels of antibiotic residues. More generally, the BST case raised the issue of "Other Legitimate Factors" inside the *Codex* Committee (see the report of the 13th Session of the *Codex* Committee on General Principles, 7-11 Sept 1998).

(non-food) sectors is left much more up to the industry. Consumers do not see things in the same light and criticise what they regard as industry's over-representation on the scientific committees of standards bodies, such as the Joint FAO/WHO Expert Committee on Food Additives, the Joint FAO/WHO Meeting on Pesticide Residues and the *Codex* committees. The "technological justifications" which allow manufacturers to use non-classified additives is another source of consumer concern. The same goes for the "codes of good practice" which manufacturers may invoke when introducing additives.[7] For example, consumers argue that substances used by manufacturers to preserve the colour of perishable foods (hydroxyquinone, polyphosphates) do not improve the products and may even be misleading as to their freshness. This highlights the difficulties of finding the right mix between highly detailed and restrictive regulations and consumers' concern at the latitude accorded to manufacturers (Doussin, 1998). It affects trade issues because the "right mix" is not seen as being the same in all countries, an issue raised in numerous disputes on the list of permitted food additives (Vogel, 1995). Much of the concern results from the introduction of new technologies in food production, which are not sufficiently transparent nor explained to consumers.

Cheese made from unpasteurised milk. The case of cheese made from unpasteurised milk provides an illustration of the fundamental differences that exist with regard to food safety between OECD countries. Regulations vary depending on different climatic and epizootic situations, different production and processing methods, and on different approaches to ensuring food safety. For example, cheeses made from unpasteurised milk are widespread in the traditional countries of cheese specialities and countries with small-scale production lines and a large variety of different types of cheeses (e.g. France, Italy, Switzerland). Other countries tend to pasteurise not only milk for direct consumption but all milk used in consumption. There were also problems within the European Union (see Mazé *et al.,* 1996) because the free movement of goods means that any country's products must be accepted in all the other member States but the EU has since harmonised rules on soft cheeses from unpasteurised milk. The European Union has been requesting international institutions, namely *Codex Alimentarius*, to recognise that good hygienic practices can ensure an appropriate level of food safety without the pasteurisation process. A 1997 *Codex* Committee on Food Hygiene report

---

7.   Codes of good farming, veterinary and technological practice play an important part in the framing of international standards. For example, a limit on veterinary drug or pesticide residues depends as a priority not on health aspects but on "good farming practice". Residues are determined with reference to such practice and not compared with the substance's toxicological credit until afterwards (Doussin, 1998).

mentions pasteurisation as one possible method to ensure the safety of milk products.[8]

Irradiated products. Since 1980, the International Atomic Energy Agency and the World Health Organisation have concluded that irradiated food (often called "ionised") presents no toxicological risk. Thirty-seven countries have approved a total of forty products that may be irradiated. Consumers in some countries, however, remain opposed to irradiation, which is relatively little used as a result (except for specific products such as spices, onions and some poultry in some countries). Here again, dissimilar consumer preferences can have an impact on trade, as for example if one country requires ground meats offered for sale on its territory to be irradiated and another refuses to use the technology. This example shows that, despite internationally accepted scientific evidence, domestic considerations can lead to different regulations and have an indirect effect on trade.

Authenticity and restrictions on production methods. Some countries, especially in Europe, have developed regulations governing the use of some geographical names, to protect regional production and improve marketing, because the identification by consumers of some specific food products has developed with reference to these geographical regions. In order to guarantee consumers the authenticity of these products, or in order to safeguard the typical character attached to the traditional production and processing methods, public regulations impose technical specifications on producers wishing to benefit from such geographical labelling. Some of these products, because of their quality or their typical character, have indeed developed over time a reputation that has led to a strong price differentiation and the creation of "market niches" (e.g. Bordeaux, Champagne, Porto, Jerez, Parma Ham). These geographical labels do not necessarily correspond to patented know-how but to a particular practice that the government wants to regulate to achieve the above mentioned objectives. Such technical specifications usually lead to higher costs of production. If producers choose to follow them rather than other production or processing methods, it is with a

---

8. The report of the Thirtieth Session of the *Codex* Committee on Food Hygiene 20-24 October 1997, with respect to Hygiene Provisions of the *Draft Revised Standards for Butter, Milkfat Products, Evaporated Milks, Sweetened Condensed Milks, Milk and Cream Powders, Cheese and Whey Cheese* and *Draft Standard for Cheeses in Brine*, includes the recommendation, "From raw material production to the point of consumption, the products covered by this standard should be subject to a combination of control measures, which may include, for example, pasteurisation, and these should be shown to achieve the appropriate level of public health protection.

view to the market premiums obtained for such "appellation" products. Problems of intellectual property rights to a region's traditional products, which without bringing novelty to the product demonstrate specific know-how, remain unsettled.[9] Wine appellations, for example, continue to be an area of contention between the European Union and the United States (Chen, 1997, Mahé and Ortalo-Magné, 1998).

Different positions also exist on the issue of restricting inputs. It is sometimes difficult to judge whether products made using other methods are fraudulent imitations. It is not always evident that the technical restrictions in question are necessary in order to obtain a particular quality.[10] Imposing such restrictions on other countries can create obstacles for their exporters. Conversely, imposing them on domestic producers alone can leave them at a competitive disadvantage.[11] The difficulties that arose when the European Union attempted to harmonise its definitions and input restrictions (e.g. in the case of beer, there are different processing standards: the German "Reinheitsgebot" dating from the XVII century, rules out the use of products other than water, malt, yeast and hops, whereas other States authorise additives like sweeteners) illustrate the complexity of such matters (Vogel, 1995). Union-wide standards were mainly established on the basis of the principle of mutual recognition. Beers circulate freely within the EU internal market, although they are brewed according to different national legislation. This, however, is more difficult to apply at the world level, since the WTO cannot establish standards itself and since there are many more countries with different regulations.

---

9.   Such concerns have been brought up to the WTO. The TRIPS Agreement (Trade-Related Aspects of Intellectual Property Rights), which came into effect on January 1995, includes geographical indications (Section 3, Article 22-24) in the areas of intellectual property rights.

10.   One example is restrictions on the shape of cheese (e.g. round cheese) which protects a tradition rather than any technical quality imperative.

11.   Some foreign wines are produced using techniques that are not accepted by EU regulations (though at present they may be imported under the terms of temporary exceptions). Techniques that are not authorised in some European countries, such as adding oak chips to wine instead of ageing in barrels, considerably reduce the production costs of wines from third countries, thus creating trade tensions. Countries that prohibit such practices argue that they have an adverse effect on quality, whereas others regard such as merely protection of traditions that bar the way to innovation.

Genetically modified organisms. A recent survey (Hoban, 1997) measured consumer acceptance of genetically modified organisms (GMO) in nineteen countries. Consumer acceptance in the United States seems to be much higher than in the countries of northern Europe, where the vast majority of consumers believe that such products represent a potential risk. Very small proportions of German and Austrian consumers seem willing to buy genetically modified products (22 per cent in Austria compared with 74 per cent in the United States). Apparently this reluctance cannot be completely explained by a lack of information about biotechnologies, since the same survey reported that a higher proportion of consumers than in the United States said they had read or heard information about biotechnologies. Part of the explanation seems to lie in cultural factors. Cultural and religious reasons were also give in other countries. Part of the explanation may also lie in the current atmosphere of mistrust surrounding health and safety systems in Europe, resulting from recent food-related health scares in a number of Member states. Consumers in Europe appear less willing to trust the assurances of public authorities and are therefore more concerned by the scientific uncertainty about the medium and long-term effects of GMOs. The practical effects of consumer reluctance on regulation and trade are as yet unclear. However, the case of genetically modified organisms -- in which the commercial stakes are considerable -- is an illustration of the impact that consumers' cultural values can have on trade.

Plant and animal health. It is in the area of plant and animal health protection that there may be the greatest number of sanitary and phytosanitary measures with potential to create barriers to trade. Often the measures taken to protect plant and animal health are quite broad (e.g. total bans on importation including from countries not contaminated, required treatments that significantly alter the product while eliminating the intended parasite or germ, measures that cover a wide range of similar products). Many of these measures would appear contestable, especially where measures are taken to prevent the introduction of a disease or parasite when it has not been proven that they already exist in that country. Moreover, the measures taken are often disproportional to the plant and animal health risk involved. The WTO Dispute Settlement Body recently dealt with a complaint from the United States relating to Japan's import prohibition of eight agricultural products (apples, cherries, peaches, walnuts, apricots, pears, plums and quince) on the ground that they are a potential hosts of codling moths, a pest of quarantine significance to Japan. A panel was established in 1997 to examine the import prohibition imposed by Japan on each variety of a product requiring quarantine treatment, even if the treatment has proved to be effective for other varieties of the same agricultural product. The panel found that Japanese measures were inconsistent with several provisions of the SPS Agreement. In November 1998, Japan notified its intention to appeal certain issues of legal interpretations developed in the panel report. The Appellate Body circulated its report in February 1999 and upheld the basic findings of the panel report.

Animal welfare. Animal welfare regulations, introduced under pressure from animal rights activists, may have important consequences for international trade. In some cases, imports of products that do not comply with certain rules may be prohibited. The European Union, for example, has banned imports of furs of thirteen types of animals that may be caught using leghold traps, even though not all of them are listed by the International Convention on Trade in Endangered Species and some species are farmed commercially (the ban only applies to countries that have not banned leghold traps). A US law designed to protect endangered sea turtles required nations catching wild shrimp and exporting them to the US be certified as having adopted conservation measures. The measure was declared an arbitrary and unjustifiable discriminatory measure to trade by the WTO dispute panel.[12] National regulations may also distort competition. The European Union has adopted directives banning the battery farming (*i.e.* rearing in individual boxes) of milk-fed veal calves and imposed collective rearing including cellulose feedstuffs which considerably increases production costs. This measure, still in the transitional phase, has little impact on trade, since there is little international trade in veal. Similar measures have been adopted for the egg sector and other measures are being discussed for the poultry meat sector, which could have a very considerable impact on the competitiveness of European poultry and egg producers, not only on export markets but also within the EU, if third countries do not adopt similar measures.[13]

---

12. In October 1996, India, Malaysia, Pakistan and Thailand took a joint complaint to the WTO in respect of a ban on importation of shrimp and shrimp products from these countries imposed by the United States. The ban was imposed on the basis that these countries did not use turtle excluder devices in shrimp nets. The complaint alleged violation of GATT Articles (elimination of quantitative restrictions) as well as nullification and impairment of benefits. The DSB Panel found that the import ban was inconsistent with GATT Articles. The US appealed the Panel decision and, while reversing some of the Panel's findings, concluded that the US measures were not justified because they were found to violate the requirements of the introductory chapter of Article XX, which states that the measures shall not be applied in a manner that constitutes unjustifiable and arbitrary discrimination. The US has indicated that it is committed to implementing the DSB recommendations by December 1999.

13. The EU Directive (1999/C 157/08) has established minimum standards for the protection of laying hens, including minimal cage dimensions, adequate perching facilities and lighting requirements. These regulations will likely

Labelling.   Government regulations on product labelling are theoretically covered by the TBT Agreement, but in practice a number of problems arise at an international level with regard to transparency, mutual recognition and control and these problems proliferate as countries impose their own specifications and labels. This is the case in particular for environmental labelling, an area of considerable disagreement. There are diverging opinions, for example, on the relevance of labels on clothing certifying low levels of pesticide use in the production of cotton, or on the specifications for labels certifying that wood products do not harm tropical rainforests. Mutual recognition of labelling for organically farmed products is difficult to achieve because countries apply the relevant criteria more or less strictly, or because some countries are considering granting such labels to genetically engineered or irradiated products. *Codex* has recently developed initial guidelines for the official harmonisation of the requirements for organic products in terms of production and marketing standards, inspection arrangements and labelling requirements while recognising that implementation is still very limited and that organic production may differ in important provisions from region to region (*Codex Alimentarius* Commission, 1999). The compatibility of environmental labels with world trade rules is uncertain and subject to dispute. The effects on trade may be considerable in a sector such as the wood industry (part of the European furniture industry has signed a charter and is contemplating using only wood that has been granted an environmental label). Labelling and consumer information policies are often portrayed as preferable alternatives to regulation because they are cheaper for producers, leave the choice to consumers and are less likely to constitute trade barriers (OECD, 1997). However, the need for international harmonisation/recognition of labels and of the underlying certification procedures raises difficulties that are comparable to the ones raised by the harmonisation/recognition of mandatory standards.

3.2.   Conceptual issues

The jurisprudence of international agreements. The Uruguay Round Agreement enabled considerable advances to be made in the regulatory sphere, for several reasons. First, it provided a more effective framework for solving disputes, through the WTO's Dispute Settlement Body. Second, it tackled the problem of non-tariff barriers through the SPS Agreement and a strengthened TBT Agreement. Third, it gave greater importance to international standards

---

increase European poultry production costs compared to countries not under the same obligations.   As a result trade flows could be affected (OECD, 1998a).

bodies. The largest number of complaints brought before the WTO in the three years since the agreement came into force have concerned sanitary and phytosanitary matters.[14] Although only a small number of disputes have been settled to date, the agreement is generally considered to have been broadly satisfactory (Hudec, 1997). Many examples may be given of countries that have changed their restrictive practices, and the very large number of notifications and discussions on potential flashpoints have made it possible to solve problems without embarking on the dispute settlement procedure (Roberts, 1997). However, as is common with international negotiations, the 1994 Agreements are framed in relatively vague terms and there is still considerable room for interpretation, particularly with respect to the sensitive issues of national sovereignty and domestic rules. Although the WTO is not based on precedent, the accumulation of a body of jurisprudence relating to the settlement of disputes will determine how they should be interpreted.

As regards the environment, a question mark remains as to the practical scope of Article XX of GATT, which mentions the protection of resources as potentially acceptable grounds for commercial discrimination, and of the TBT Agreement, which mentions it as possible grounds for exceptions to international standards. Opinions on the subject differ (Gray *et al.*, 1996).[15] More generally, Rege (1994) points out a large number of contradictions between the different agreements and emphasises the need for a "collective interpretation" of them, or even an amendment to Article XX, so as to recognise the role of multilateral agreements on the environment. There are also a number of other regional and multilateral environmental agreements, and many ambiguities remain concerning the respective roles of United Nations agreements, regional agreements and the WTO (Mahé, 1997).

14.	Seventy-three complaints were brought under the dispute settlement procedure during the first three years of the agreement, of which 17 concerned agricultural products and 7 related to sanitary and phytosanitary aspects. The complaints were as follows: two (from the United States and Canada) against the European Union concerning cattle hormones; two against Australia concerning salmon imports; one against South Korea concerning the inspection of agricultural products; one against the United States concerning poultry imports; one against Japan concerning quarantine measures; one against South Korea concerning the sell-by date of food products.

15.	Articles XX and XXI allow exceptions to the GATT for a list of clearly stated public policy concerns, including health, the environment and national security.

A number of disputes involving several OECD countries have already been brought before the WTO since the dispute settlement procedure was established in 1995, though a number of conflicts have been resolved on a bilateral basis in the desire to avoid establishment of a formal panel process. Only three panel reports have been released to date in the context of the SPS Agreement (EU/US/Canada hormone treated beef, Canada/Australia salmon imports, US/Japan quarantine regulations) but these WTO cases have helped to clarify some provisions of the SPS Agreement and to a certain extent, they have provided some guidance to governments for the conception and implementation of their SPS policy. In its conclusions, the 1997 beef/hormones panel attributed a key role to the *Codex Alimentarius*. The *Codex* thus imposed itself as a set of standards prevailing over national regulations. The appeal judgement of January 1998 overturned some of the panel's conclusions.[16] Present case law would therefore seem to support the principle that the country, which considers a measure to be unlawful, must prove that to be so. However, these judgements do not exhaust the various questions and problems that can arise from the implementation of the SPS Agreement nor are they necessarily applicable to other conflicts, each of which must be arbitrated by the WTO on a case-by-case basis. The economic stakes are high and such disputes are likely to remain a priority in the future trade agenda.

Lastly, a certain number of points of friction between OECD countries concerning regional appellations and patents have not yet been settled. Intellectual property rights to biological resources remain a grey area. There is as yet no multilateral framework within which emerging controversies over patents on living creatures can be discussed, points of view harmonised and clear rules drawn up. The various agreements establish a framework for regulation but not for operating rules (Mahé, 1997, 1998).

---

16. In a nutshell, the conclusions of the appeal judges (January 1998) are as follows. First, any WTO member is entitled to set a higher level of consumer protection than the level recommended by the *Codex Alimentarius*, provided that it is scientifically justified. Second, a qualified scientific opinion may be taken into consideration even if it is a minority opinion. Third, the EU's ban was not based on adequate risk assessment and it should therefore rectify the situation. Fourth, in future it will be for the plaintiff to prove that a measure taken by a member of the Organisation is incompatible with the SPS agreement (the panel had recommended that the onus of proof should lie with the defendant). It is worth noting that the conclusions of the Appellate Body stipulated that risk assessment does not have to be quantitative in nature, does not have to come to monolithic conclusions reflecting the main-stream scientific opinion and that it can include other factors such as good veterinary practice.

Different conceptions of risk. It is widely recognised that it is seldom possible nor economically feasible to achieve zero risk with respect to food safety. The SPS Agreement explicitly requires Members to base their SPS measures on risk assessment as appropriate to the circumstances of the risk to human, animal or plant health (Article 5.1).[17] However, there is no agreement on what constitutes justifiable risk or "acceptable risk" as mentioned in the SPS Agreement (Annex A.5). Nor is there any agreement on the importance to be given to risk analysis, or on what is meant by the term "risk", or on methodology. Within the OECD, approaches may differ widely from one country to another, especially concerning the importance to be placed on risk management (Mazurek, 1996). Some countries prefer to emphasise risk elimination (e.g. sterilisation of mineral waters, ban on cheese made from unpasteurised milk, etc.). Others emphasise the possibility of risk control (in the above mentioned examples by bottling at source, HACCP controls, etc.), which is less costly and alters the product less, and points to the inconsistency of seeking to achieve zero risk in one area while tolerating high risk in others.[18]

Some consumers and governments are not satisfied with some *Codex* standards. Powell (1997) highlights the difficulties of obtaining unanimous and definitive scientific assessments of the hazards present in food, because of genetic mutations, for example, or combinations of pathogens with uncertain effects, or the influence of exogenous and unforeseeable factors on micro-organisms. The standards accepted by scientists do not always have an indisputable scientific foundation: some have had to be completely revised at various times, and scientific "unanimity" is seldom achievable, especially with regard to the carcinogenic properties of products (Mazurek, 1996).[19]

---

17. Generally risk assessment is regarded as a component of risk analysis, which is a three-stage process. The first stage, risk assessment, consists in identifying hazards, in particular their forms, thresholds and probabilities. This phase is essentially scientific even though political elements, such as what is regarded as an acceptable level of protection, may impinge. The second phase is risk management and the third phase is communication concerning the risk.

18. The SPS Agreement states that countries should have the objective of "consistency". If, on the basis of a risk assessment, there is a one in a million chance of a certain product causing a certain level of damage, the product should not be subject to greater restrictions than other products presenting a similar level of risk. The level of risk may be acceptable or not, the objective is that the acceptable risk should not be different according to the product concerned.

19. Standards concerning dioxin, for example, which may be found in dairy products and accumulates in organisms and in mother's milk, have been

International standards are now put to the vote at the Codex, and some are passed by a small majority. Not all countries are willing to acknowledge the legitimacy of risk thresholds imposed on them in this way.[20]

Consumers assert the right to entertain fears that some scientists regard as irrational, especially concerning genetically modified organisms and irradiated food. Controlling short-term risks does not mean that long-term risks or uncertainties do not exist. Environmental and consumer groups argue for the inclusion in the SPS and TBT Agreements (or more generally in the WTO Agreement) of a "precautionary principle", which would allow exceptions to the regulations in cases where scientific evidence is insufficient or there is scientific uncertainty (Rege, 1994). According to the precautionary principle, precautionary measures should be taken in this absence of certainty about the state of scientific knowledge at the time (Godard, 1997).

The precautionary principle is recognised in several international agreements (e.g. International Convention on the Protection of the North Sea, Rio Declaration, Framework Convention on Climate Change), Principle 15 of the Declaration of the United Nations Conference on the Environment and Development, adopted in 1992 in European law (Maastricht Treaty) and in national law. According to the Rio Declaration, "in *order to protect the environment, the precautionary approach shall be applied by a State, according to its capabilities. Where there are threats of serious or irreversible damages, lack of full scientific certainty shall not be used as a reason for postponing cost effective measures to prevent environmental degradation"*. Article 5.7 of the SPS Agreement indicates that if relevant scientific evidence is "insufficient", members may adopt SPS measures, on a provisional basis, while seeking additional information about the risks posed by a hazard.[21] In addition,

---

established on the basis of experiments on rats extrapolated to humans. Even with a considerable safety margin, the basis for such standards seems relatively arbitrary. According to Antle, US Environment Protection Agency estimates for cancer risk from pesticide residues are approximately 1000 times higher than equivalent risk estimates using other methods (Antle, 1995:5).

20. The revision of the acceptance procedure is presently being discussed within the *Codex* Committee on general principles, where new voting modalities relying on a larger majority were proposed in September 1998.

21. SPS Agreement Article 5.7: In cases where relevant scientific evidence is insufficient, a Member may provisionally adopt sanitary or phytosanitary measures on the basis of available pertinent information, including that from the relevant international organizations as well as from sanitary or phytosanitary measures applied by other Members. In such circumstances,

the Appellate body of the hormone in beef case stated that the precautionary principle is not limited to Article 5.7 but draws also upon Article 3.3 and the Preamble. However, provisions of the SPS Agreement regarding precaution, are much more restrictive than what some consumer groups often mean when they invoke the "precautionary principle", suggesting that there may be a fundamental ambiguity between the expectations of certain groups in society and practical measures.

Disagreement on quality attributes. There is considerable disagreement on quality attributes, such as the nutritional content and production methods, and on the extent to which they may legitimately be the subject of international regulation. Some countries consider that the soil, climate and traditional know-how that exist in a region have a decisive influence on product quality, others do not. These notions of product quality are ill matched to the more restrictive approach adopted *de facto* internationally.[22] The stance of the SPS Agreement is to take into consideration only sanitary and phytosanitary factors. International Standardisation Office (ISO) labels, which could become *de facto* standards regulating international trade, do not include all the quality dimensions of European regulations, which are based to a considerable extent on a product's organoleptic qualities (taste) and authenticity (Sylvander, 1996).[23]

The EU has regulated the use of some geographical names that are attached to specific production, in particular appellations of origin, *inter alia* because of domestic considerations about the defence of quality or specificity of these products. However, there are certain difficulties when it comes to the international protection of these names for trade purpose. A number of trade disputes arise from what is considered by one party as usurpation of brand names and by others as generic or semi-generic names. Bilateral agreements

---

Members shall seek to obtain the additional information necessary for a more objective assessment of risk and review the sanitary or phytosanitary measure accordingly within a reasonable period of time.

22. Chen (1997) highlights the incompatibility of European quality marks, which emphasise authenticity, with US legislation, and the difficulty of achieving international recognition for this type of mark.

23. The International Standardisation Office defines standards for the sole purpose of facilitating trade. Unlike bodies such as the *Codex* or the IPPC, the members are generally represented not by government bodes but by industry bodies. The ISO acts as a third party, offering independent quality certification. The ISO system is voluntary, but there is growing pressure for the ISO 9000 system to be incorporated into the procedures of the Codex.

and mutual recognition is one way of solving such disputes. At multilateral level, the TRIPs agreement is the framework that should help address these issues (Annex 5).

There is also disagreement on ethical and cultural quality attributes. Growing numbers of consumers are concerned about the possible adverse effects of their purchases, on the destruction of natural resources in other countries, for example, or on child labour.[24] There is growing pressure from public opinion for the imposition of more environment-friendly practices in third countries, especially in order to protect the "common resources of humanity" such as tropical rainforest. Consumers are also concerned about the importation of goods which they reject for cultural or religious reasons. But the fact that consumers' ethical values are not the same in all countries may generate trade tensions. In part due to the pressure of public opinion, some countries have introduced regulations in areas such as animal welfare, the protection of fauna and flora and genetically modified organisms, which may considerably increase the cost of producing food. The SPS and TBT Agreements can put countries in an uncomfortable position by requiring them to authorise imports of goods produced using methods which they have decided to ban at home. The European Union, for example, argues that it should not be obliged to authorise imports of food produced under less restrictive livestock farming conditions than its own (e.g. animal welfare), or using biotechnologies that consumer pressure prevents its own farmers from using (e.g. BST). Even though these problems are not unique to the food industry (they have been around for many years in the sphere of social and environmental legislation), the impact on trade is an important issue for the future (see Baghwati and Hudec, 1997).

The difficult compromise between consumers' legitimate aspirations and fair trade. New consumer values may well be out of step with GATT principles. In the environmental sphere, rulings in disputes brought within the framework of GATT and the WTO restrict a country's ability to use trade measures restricting imports to protect natural resources outside its territory, even in the case of resources which some consider to be "common to humanity". In the cultural sphere, a country may introduce regulations that are more stringent than international standards on ethical, moral or religious grounds only under very limited conditions. The SPS Agreement does

---

24. According to some pressure groups, recent surveys have estimated that the purchasing and investment decisions of almost 50 million consumers in the United States are influenced by ethical or environmental considerations (Ray, 1996).

recognise the validity of consumer concerns to some extent under Article XX as does the TBT Agreement, for example, by authorising different labelling.[25]

Ignoring consumer values may result in a falling away of their support for the process of trade liberalisation. New media, especially the Internet, can provide a considerable soundboard for the detractors of international trade. If consumers consider the way in which children are exploited, cosmetics tested, foxes killed or cattle reared to be an integral component of the quality of a product, their demand for such products is altered by the presence or absence on the market of goods which do not comply with their ethical values. In particular, if imports are suspect according to one of the criteria described above, trade liberalisation can have an impact on overall demand. This is the case, for example, if it is difficult for consumers to identify goods produced under such conditions. The banning of GM foods by some European countries (Austria, France, Luxembourg) and by some food retailers (primarily in the UK) illustrates the potential trade impacts of the agro-food industry launching new products and technologies without taking account of consumer's health, environmental and ethical concerns.

A number of arguments can be made for including ethical and cultural values as grounds for trade restrictions. Reluctance to consume goods produced in unethical conditions can affect demand for all goods, including those produced "virtuously", as for example with animals caught in traps and animals reared on farms. Externalities between goods may arise if ethical and cultural values are acknowledged. Trade is also one of the most effective means for obliging countries to respect human or children's rights or to protect natural resources and endangered species. Socially aware consumers who would like to be able to wield such a weapon find it hard to understand why international trade rules should prevent them from doing so. Moreover, it may be paradoxical to reject trade restrictions for cultural reasons when they are admitted for non-food products such as medicines.[26] Vogel (1995) gives several

---

25.   For example, Article XX of the GATT 1994 (General Exceptions) allows import restrictions when *"necessary for the protection of public morals -- Article XXa"* (though in practice religious grounds may also be invoked), and when *"relating to the conservation of exhaustible natural resources* if such measures are made effective in conjunction with restrictions on domestic production or consumption -- Article XXg"* (interpreted as including biological resources).

26.   The case of RU 486, the "morning after" contraceptive pill, is an extreme example of a product which may not be imported into certain OECD countries, solely for cultural reasons, though many other examples exist in the medical sector.

examples where trade restrictions and a desire for access to greener markets have had an impact on a country's attempts to improve social and environmental regulation.

However, giving consideration to ethical, cultural or moral arguments could open a Pandora's box. For some countries, risk may be social as well as biological, including factors such as bankruptcy among farmers and rural desertification (Lorvellec, 1997). Cultural or ethical arguments could be used to cover a potentially unlimited number of exceptions to trade. A laxist interpretation of the WTO Agreement in this sphere would provide justification for a whole host of trade barriers. Consumer concerns about the way imported goods are produced can be used by pressure groups acting in their own interests. For that reason, trade restrictions motivated by social, cultural, ethical or environmental considerations could be a form of protectionism in disguise. There can be convergence between consumer demands for stricter standards than those recommended by scientists and the economic attraction of strengthening non-tariff barriers. This problem is likely to be an area of contention in coming years. The challenge for governments is to find the right balance between consumer protection and reducing technical barriers to trade.

4. Contributions of economic analysis

4.1. Economic analysis and quality

Socially optimum quality. It is possible to give an economic definition of socially optimum quality, which depends on the cost of producing higher quality goods and consumer willingness to pay for a given level of quality. The same is true for the specific case of product safety. If producers and consumers are perfectly informed about product characteristics, socially optimum quality can be obtained on a competitive market which allows the supply of and demand for quality free play. As far as food safety is concerned, this means that a safety level will emerge whose cost will be regarded as acceptable by consumers, who agree to pay for it (this economic conception of the "right" level of product safety differs from the widespread notion that zero risk is preferable, because additional safety can never be obtained at zero cost). The market may be imperfect, however, and the play of supply and demand may spontaneously produce an offered quality which is too low, for example. The economic mechanisms at work in such cases is described in the following sections.

International trade and product quality. In international trade theory, the question of quality is approached by considering the markets on which the products are differentiated. In classic theory, a distinction is made between two types of product differentiation, vertical and horizontal. In vertical differentiation, higher quality corresponds to market segmentation according to consumer incomes (at equal prices, all consumers would unambiguously choose higher quality). Horizontal differentiation, on the other hand, is based more on the notion of taste and corresponds to market segmentation according to individual preference (at equal prices, consumers would choose different products).

There are many arguments for trade liberalisation, and most of them remain unchallenged by recent developments in economic theory relating to differentiated products. Country specialisation according to comparative advantage leads to lower real prices and hence, for a given level of expenditure, allows the possibility of consuming the same amount but a higher quality of products (vertical differentiation). In other cases, trade liberalisation reduces the cost of the consumer's preferred basket (horizontal differentiation). Countries can take advantage of growing economies of scale when markets are opened up; this in turn reduces production costs at constant quality. The increased competition resulting from trade liberalisation encourages firms to offer a better quality/price mix. In addition, international trade increases product variety, and freedom to choose generally has a beneficial effect on consumer welfare, since welfare suffers when regulatory barriers prevent consumers from gaining access to higher quality foreign products.[27] International trade also favours the rapid spread of technology, stepping up the pace of technological progress, which in turn favours an increase in product quality at constant prices.

However, the beneficial effects of trade liberalisation may be attenuated by spontaneous market inefficiencies. In such cases, government intervention can be modified in order to avoid negative effects and to maximise

---

27.   In the case of vertical differentiation, regulatory barriers to trade typically result in a restriction on consumer choice in the available quality segment. The market may not be covered as a result, and a significant fringe population may be deprived of the opportunity to consume a cheap but lower quality product that it wishes to acquire (Bureau *et al.*, 1997 show this mechanism at work in the case of the European Union ban on hormone-treated meat). In the case of horizontal differentiation, trade restrictions take the form of less variety in the products on the market. This also results in a loss of consumer welfare, insofar as the choice between a large number of products is a positive element in their utility function (see Krugman, 1979).

the gains from trade. The following sections focus on three issues in connection with the question of trade liberalisation and food quality -- imperfect competition, imperfect knowledge and the presence of risk.

4.2. *Imperfect competition and less than optimum quality*

Trade liberalisation generally results in a challenge to oligopolistic situations and favours greater competition. However, the links between competition and product quality are ambiguous. A monopoly firm will seek to limit the quantity on offer but may in some cases set the quality on offer at an undesirable level for the firm in order to maximise its private profit.[28] By increasing competition, trade liberalisation will have a positive influence on the quality on offer. But greater competition on a market may also have perverse effects. Greater competition may result in certain quality segments not being supplied at all, or in an overall decline in quality on the market (Gal-Or, 1989a). When firms have fixed costs, it has also been shown that greater competition can cause producers to set quality levels further removed from the socially optimum level in order to limit the decline in their profits (Reitzes, 1992).

All else being equal, trade liberalisation, by increasing competition in industry, could encourage quality-cutting wars or fictitious differentiation, especially if consumers track prices more readily than quality (Beales *et al*, 1981). When there are negative links between quantitative output and quality, greater competition could encourage firms to offer lower quality products. In practice, it is difficult to imagine that any adverse effects on quality caused by "excessive" competition leading to sub-optimal product differentiation, as a result of the opening up of frontiers, would outweigh the benefits of trade liberalisation. Provided that consumers are not being misled when a quality is claimed, quality cutting competition is unlikely to be a major empirical problem. Nevertheless, it is a point on which theory remains ambiguous, and cases where competition from foreign firms has an adverse effect on marketing, development and availability of quality products cannot be ruled out. Various concerns have been raised about differences in standards,

---

28. Basically, this works as follows: social welfare depends on the sum of consumers' willingness to pay, whereas a monopoly setting the quality of its output considers only the marginal consumer's willingness to pay. If the two are very different, the quality choice will be less than optimum from a social standpoint (Spence, 1975;). More generally, in a situation of imperfect competition there are few reasons why producers setting both the quality and quantity levels of their products simultaneously should offer socially optimum quality (Krouse, 1990).

and more particularly on labelling of products (e.g. EU internal disputes on the definition of chocolate and yoghurt). If harmonisation or mutual recognition of standards was to result in the lowest common denominator, market globalisation could result in the loss of some segments of quality products.

4.3.   *Imperfect information*

Broadly speaking, if consumers are not fully informed about product characteristics they may consume a dangerous product, or acquire a quality they do not wish to consume, or pay a price that does not reflect the real quality of the good in question. In all these cases, the level of welfare in society may be lower than it would have been if information had been perfect. The workings of the market may cause vendors to offer an inadequate level of quality or safety when information is imperfect. In some countries, the threat of individual or class legal actions of consumers against the industry plays a significant role in regulating industry behaviour (e.g. Anglo-Saxon countries, in particular the United States). In others, where legal prosecution or private liability are not developed to the same extent, there may be a greater need for public regulation to protect consumers.

Consumer goods may be divided into search, experience and credence goods. A good is a "search" good when the consumer is capable of assessing its quality before buying it, an "experience" good when the consumer discovers the quality only after consuming it, and a "credence" good when the consumer never discovers the quality of the good (or does so only in the very long term). Search goods are goods for which perfect information can be acquired, albeit at a cost. The organoleptic components of quality (taste) generally fall into the "experience" category. But many agro-food goods fall into the "credence" category (Caswell and Mojduska, 1996). This is the case, for example, when the "safety" component of quality or the nutritional composition of a product are at issue. It is also the case with the ethical, cultural or environmental components of quality. The economic mechanisms at work in these three categories are different. With experience goods, for example, the incentives for quality fraud are limited by consumer sanctions on the occasion of repeat purchases. With credence goods, there is no spontaneous mechanism for market regulation and it is more difficult to indicate quality in a credible way. The market failures highlighted by Akerlof may extend into the long term (Box 1).

The market dysfunctions caused by imperfect information described in Box 1 may reduce the beneficial effects of trade liberalisation for general welfare. Opening up markets can result in the coexistence alongside local products of foreign products whose quality is less familiar to domestic consumers. The imported goods may be perceived as being of lower quality because of doubts as to foreign control procedures or the different importance attached to each component of the overall quality of the good. By affecting consumer perceptions about the average quality of the products on the market, trade liberalisation may therefore contribute to a reduction in demand or in consumers' willingness to pay.

With experience goods, although foreign producers have a cost advantage for low quality goods, opening up frontiers alters the price distortion needed to signal the high quality of domestic products. Trade liberalisation may reduce prices in low quality product segments while increasing prices in high quality product segments. The second effect has a negative impact on consumer welfare which may offset the positive impact of the first (this argument is developed by Falvey, 1989). It is difficult to monitor the production process of imported credence goods, which is the sole means for acquiring information about their quality. Foreign firms are also less exposed to judicial sanctions (liability), which may encourage fraud when the consumer is unable to verify the quality of the good in question directly. This also induces welfare costs which limit the beneficial effects of trade liberalisation.

In some cases, it is conceivable that more information could reduce sales by making consumers aware that a certain practice exists (typically, one of the reasons that few people object to irradiated or genetically modified food in some countries, is that they do not realise that products they consume are affected). When consumers know that they do not have complete information, markets are affected in a way that reduce transactions. When information is imperfect, trade liberalisation can encourage phenomena such as the elimination of high quality products or price distortions, especially where credence goods are concerned. Consumer uncertainty as to the type of products on the market (which might result, for example, from imports of goods like hormone-treated meat or genetically modified seeds) could affect demand and raise this sort of problem (Bureau *et al.*, 1997). It is theoretically possible for the welfare loss resulting from reduced consumer willingness to pay to outweigh the welfare gain resulting from cheaper imports. Under these circumstances, it is possible for trade liberalisation to result in greater trade flows but a decline in collective welfare world-wide (Thilmany and Barrett, 1997; Bureau *et al.*, 1997). It remains to be proved whether what is possible in theory will actually happen in practice. At a theoretical level, however, there is a substantial body of research to show that information effects may limit the welfare gains made possible by trade liberalisation as a result of lower prices and greater product diversity (Falvey, 1989; Grossman and Shapiro, 1988).

Box 1. Imperfect information and market inefficiency

Imperfect information about product quality can cause market dysfunctions affecting the price and quality of traded goods. Two major types of problems are generally identified. In the case of adverse selection, vendors do not decide the quality of their products. In the case of moral hazard, vendors do decide the quality of their products. With regard to food safety, for example, long-term risks and risks over which producers have no control fall into the adverse selection category, while risks over which they do have control fall into the moral hazard category.

When consumers are not able to distinguish the specific quality of different products, they are not willing to pay as high a price as they would if they were sure that the product was of high quality. Akerlof (1970) has shown that imperfect consumer information about product quality could even result in total close-down of the market (absence of trade) if, because of a lack of information, buyers' willingness to pay was insufficient to cover production costs. If buyers' willingness to pay is less than the cost of producing high quality goods, only low quality goods (less costly to produce) are traded and high quality is frozen out of the market (Akerlof quoted second-hand cars as a famous example of poor quality chasing away high quality).

In order to alleviate these dysfunctions, vendors may signal the quality of their products. For an **experience good**, this signal may be communicated through the price. In the case of moral hazard, Shapiro (1983) has shown that a higher price than the perfect information price could encourage producers to offer high quality on a lasting basis. This price supplement constitutes the information rent, which enables quality to be maintained over time and creates an incentive not to cheat on quality.[29] This can be a way of segmenting the market and informing consumers. However, there is a cost to society in comparison with a situation of perfect information, since consumers have to pay the higher price needed to signal quality. This mechanism does not work with **credence goods**. As consumers never detect the quality of the product, repeat purchases do not bring them any additional information and will not change their behaviour, thus not providing any incentive to producers to offer high quality.

The price mechanism described above is not the only way in which quality can be signalled. Firms can spend heavily on advertising, for example, and consumers will anticipate that such expenditure can only be covered by an information rent, indicating that the product is of high quality (if it is an experience good). Warranties or substantial compensation in the event of problems are other types of signal (Gal-Or, 1989b, Lutz, 1989). Firms can also secure a reputation via a trademark or a quality signal such as a label. Governments can enhance the credibility of quality signals by certifying them. Conversely, consumers may use intermediaries such as brokers or dealers who, better informed, can give them credible information on the quality of the offered goods. In all these cases, however, imperfect information has a welfare cost to society.

---

29.   A high quality producer has an economic interest in preserving his reputation. If he sold a poor quality product with a high quality price ticket, buyers detecting the quality after consuming the product would not buy from the same producer again. The cost of establishing a reputation, which can be amortised only by the information rent, acts as an incentive and prevents low quality producers from signalling their products as high quality products.

4.4. *Economic aspects of increased risk resulting from international trade*

By allowing the spread of pests and pathogens, international trade may increase sanitary and phytosanitary risk. The blocking of sewers in North American cities by exotic molluscs brought in on the hulls of cargo ships is one example of this phenomenon. Various outbreaks of food poisoning in the United States caused by imported foods provide another example (e.g. gastro-enteritis -- *Cyclospora* -- apparently due to fruit from Guatemala, epidemics linked to Peruvian carrots, Mexican strawberries, Thai coconut milk, Chinese mushrooms, Israeli snack foods, etc.; see *New York Times*, 22 September 1997). This imposes costs on society, such as lower yields in the case of an epizootic or the spread of a plant pest, or healthcare costs. These costs may be considerable because of the externalities that exist where epidemics are concerned (although imported products were not necessarily at issue, the costs of the epizootics of foot and mouth disease in Taipei in 1997 and of swine fever in the Netherlands in 1996 were considerable; OECD, 1998a).

Trade liberalisation therefore modifies the "public good" aspects of food safety. For example, it can be a country's reputation as a producer of safe food which suffers when there is an outbreak of a food-borne disease, affecting all exporters through the misdeed of only one producer. By increasing the number of consumers of a given product, freer trade may also modify the economically optimal provision of food safety, since, as it is generally the case with public goods, this provision is a function of the sum of the willingness to pay of all consumers (Henry, 1991).

Usually, when economic agents are risk-averse they are willing to pay in order to eliminate risk (e.g. in the form of insurance), and the social cost of risk may be estimated as the difference between the anticipated welfare and the welfare of society that would prevail in a situation without uncertainty.[30] As far as food safety is concerned, however, consumer aversion to risk is complex. In some cases, consumers are willing to pay a very high "premium" for zero risk in comparison with very low risk (Magat and Viscusi, 1993). In addition, consumers' behaviour can be affected by what they imagine to be a risk, even if

---

30. A simple example is that of production in an uncertain world: in order to continue production in the presence of climate risk or price risk, firms demand a premium which is reflected in higher production prices. The price is paid by consumers, whose welfare is reduced by a corresponding amount. In comparison with a risk-free situation, the producers' welfare is unchanged but the consumers' welfare has declined.

it is not backed up by scientific evidence, resulting in costs to society.[31] Trade liberalisation, which exposes consumers to less familiar products than the domestic ones, may increase the number of cases where consumers have an inaccurate evaluation of the risk level, raising some very difficult questions for economists. Pollak (1995, 1998) shows that the way imagined risks should be accounted for by economists, in particular in cost-benefit analysis, is still unclear.

It is not easy to assess the extent to which any increase in sanitary and phytosanitary risk is attributable to international trade rather than other factors such as tourism, and the previous sections show that estimating the social costs of these extra risks is difficult. However, the increased risk may limit the welfare gains from international trade and should, in any rigorous approach, be taken into consideration when assessing the gains resulting from trade liberalisation. On this point, it is worth noting that an assessment of sanitary risk alone, provides no basis whatsoever for deducing the cost of that risk to society. Only one product in a thousand may present a risk, but many more than one in a thousand consumers may refuse to buy the product. In such cases, risk analysis needs to include economic aspects. The SPS Agreement distinguishes between public health aspects and animal and plant health. Only for the latter is it foreseen that "relevant economic factors" should be taken into account.

4.5. *The role and place of economic analysis*

Cost-benefit analysis. If food safety or quality regulations are ineffective, consumers do not obtain the level of quality for which they are willing to pay, and the quality to which they have access is not offered to them

---

31. Magat and Viscusi (1993) have observed many cases where consumers seem to deviate from "rational" behaviour, inasmuch as their behaviour was inconsistent with the predictions of the anticipated subjective utility model. According to the authors, individuals accord excessive importance to low probabilities of risk. The questions raised by risk misperception are complex. Even when the public fears are not shared by the experts, these fears affect their economic behaviour and therefore, the state of an economy. Governmental regulations should therefore account for unjustified public fears to a certain extent. In addition, in a democratic society, the governement ought to respond to consumers' worries, however remote and conservatively estimated the risk (Margolis, 1996). However, spending money on expensive treatment of (clean) water, just because "people feel worried about some risk" would be clearly economically inefficient (Pollak, 1998).

at the minimum price. A regulation should therefore be assessed in the light of the benefits it brings in comparison with the absence of regulation and the costs it imposes on the economy.

Cost-benefit analysis is used to enable public authorities to take decisions concerning national regulations. Arrow *et al,* (1996), for example, recommend that the method should be used systematically, since they observed very considerable differences between the cost of public health measures and their real impact on health (they give estimates where, within the same agency, the cost per life saved varies between US$200 000 and US$10 000 000 depending on the programme, which means that more lives could be saved at the same cost to society). Even though society does not accept all risks in the same way, and even though social choices cannot be reduced to the equalisation of a statistical cost between programmes, cost benefit analysis is an important stage in the framing of regulations. It is mandatory for projects of a certain size in the healthcare sector in the United States (*Executive Order* 12291, 1981).

Methods based on estimates of the cost of illness or the cost of shortened human life may also be used to assess the benefit of regulations (Box 2). These costs can then be compared with the costs of sanitary regulations (Viscusi, 1993). In practice, this raises many technical and methodological problems. For example, estimates of cancer risk from pesticide residues contain a substantial degree of uncertainty as to the risk, making any economic estimate particularly difficult (Antle, 1995; Mazurek, 1996). In addition, it is not possible to calculate the probability of a risk that is too uncertain, making it difficult to carry out analysis with conventional tools. This is the case, for example, with the risk of genetically modified organisms propagating genes, or the risk of long-term epidemics such as BSE and CJD, or environmental risks. Measuring the benefits procured by regulations designed to guarantee certain ethical or cultural aspects of product quality is no easy matter either, and the problem of the valuation of imagined risks is a difficult one. However, it is possible to use approaches based on the measurement of changes in the consumer utility function when consumers have access to a product with attributes to which they are attached (Kopp *et al.,* 1997). If, for example, consumers place particular value on the fact that a good is produced without the use of biotechnology or irradiation techniques, estimating their willingness to pay means that the variation in consumer satisfaction resulting from a regulation prohibiting the technique in question can be quantified in money terms (Viscusi *et al.,* 1995; Magat and Viscusi, 1993).

Box 2. Methods for estimating the benefits of sanitary and technical regulations

Where food safety and the spread of plant and animal diseases are concerned, cost benefit analysis involves quantifying the level of risk and estimating its economic impact.

This approach is widely used, though very unevenly from one country to another, not only in order to assess the interest of a regulation but also to compare the advantages and disadvantages of several possible means of government intervention. In particular, it can be used to rationalise the strengthening of sanitary controls in relation to the dissemination of information and the raising of consumer awareness, or to inform decisions about the introduction of regulatory standards (Kopp *et al.*, 1997).

Several methods exist for estimating the cost of mortality and morbidity and evaluating in money terms the benefits of government action resulting in a reduction of sanitary risk. With the ***human capital*** method, a value is placed on the reduced risk of premature death based on an evaluation of discounted labour flow. For an individual of a given age, the value of the life prolonged (statistically) by a regulation corresponds to the discounted sum of the mathematical expectation of the person's revenues (Freeman, 1993). Some extensions of this method have been proposed, in particular by integrating non-merchant aspects and the value of the individual's descendants (Viscusi, 1993). With the ***illness cost*** method, a value is placed on the reduced morbidity resulting from sanitary or regulatory methods, based on an estimate of medical costs and productivity losses due to illness (Buzby *et al.*, 1996). Opportunity costs from investing in activities that reduce the risk are included in the value of reduced illness (Landelfeld and Seskin, 1982). As with the human capital method, statistical methods have to be used to estimate the risk, especially dose-effect relationships.

Methods based on estimates of willingness to pay, although more difficult to apply, are wider-reaching, since they make it possible to include quality-related aspects that cannot be translated into identifiable short-term illness. The ***preventive expenditure*** method seeks to measure agents' willingness to pay by observing the efforts made to avoid illness. With this method, a money evaluation of the disutility of being ill is added to the estimated cost of illness, together with an estimate of the preventive expenditure that an individual is willing to commit according to a given pathogen level (Harrington and Portney, 1987). ***Contingent evaluation*** methods involve asking individuals directly about their willingness to pay in order to reduce the risk of an illness, or more generally to obtain higher quality in a good. By directly revealing willingness to pay, this method theoretically makes it possible to gain a money estimate of all the benefits arising from a given measure. However, answers have to be corrected for statistical bias due to respondents' incentives to over- or underestimate their willingness to pay (which depends in particular on whether they anticipate having to pay the disclosed sum or not). As these methods are widely applied to environmental issues, efforts have been made recently to harmonise survey methodologies (see, in the United States, NOAA panel, *Federal Register* 58,10). Another method being used increasingly widely at present is the ***experimental economy*** method, which involves getting a group of individuals in a situation where their real behaviour is simulated to reveal their willingness to pay for particular qualities. Such methods are relatively onerous to put in place, but they make it possible to obtain a precise measurement of the value that a sample of individuals places on different sanitary thresholds, according to information received, for example (Hayes *et al.*, 1995).

(continued on next page)

International agreements on sanitary and technical measures do not oblige countries to adopt only those regulations whose benefits exceed their costs (Roberts, 1997). In practice, many countries introduce import restrictions on sanitary grounds, to avoid the spread of pests for example, without making any prior estimate of potential losses. These may sometimes be very small in comparison with the cost to consumers caused by the regulation in question. If economic methods of calculation were used more systematically, the welfare gains resulting from the import restrictions could be compared with the welfare gains resulting from freer trade.

4.6. *Economic analysis and the settlement of disputes*

Scientific and economic criteria. Present arrangements for the settlement of disputes relating to technical and sanitary barriers have put economic analysis second to risk assessment. However, economic analyses can help decision-makers when choosing between different risk management options and when reviewing quarantine policies (James and Anderson, 1998). It can also provide a more sound basis for discussing the role of "other legitimate factors" than health hazards, a recognised problem (e.g. see the item on "the role of science and the extent to which other factors are taken into account", 13th Session of the *Codex* Committee on General Principles, 7-11 Sept 1998, Paris).

The idea of objective science serving to guide the adoption of SPS measures while minimising their negative trade effects prevails in the SPS Agreement. In practice, economic and political considerations are very much intermingled. In many cases thresholds have been set not only on the basis of medical effects but also on the basis of what is technically and economically feasible, and many scientists acknowledge, off the record, that some standards are defined "after the event" (radioactivity thresholds, for example). Ever since scientists' recommendations acquired the status of potentially mandatory standards, with considerable economic interests at stake, it has been difficult for them to ignore economic considerations. Salter (1988), Powell (1997) and Hillman (1997) have given numerous examples of "mandated science" or "negotiated science". Manufacturers are also strongly represented on *Codex* and joint FAO/WHO committees.

A trade-off between costs and benefits is sometimes implicit behind the scientific criteria, in the form of the setting of standards which take economic factors into consideration and reference to risk analysis in the settlement of disputes. Risk analysis includes a risk management component, which is primarily based on scientific elements and refers to the ways in which risk may be reduced to an "acceptable" level. However, risk management may include economic considerations and in the last resort the decisions taken are often of a political nature. Giving cost-benefit analysis a more explicit role could be helpful for risk managers and make the process more transparent.

A parallel with competition policy. The procedure for settling sanitary and technical disputes under the auspices of the WTO could draw on the competition policy example. Virtually all OECD country competition policies are ultimately directed not at preserving and enhancing competition as an end objective. Instead they focus on greater consumer welfare as the final objective. Consequently, most competition agencies are prepared to make certain tradeoffs (Viscusi *et al*, 1995). For example, if a number of competing, smaller grocers were to agree to pool their purchasing power this could tend to reduce the degree of competition prevailing among them at least on the purchasing side. Common purchasing might nevertheless be tolerated by competition agencies if they were convinced that the arrangement reduces competition to the minimum necessary to enable smaller retailers to be more efficient hence more effective competitors for larger chain grocers. Another example could be found in the steps manufacturers might take to ensure that each of its dealers sells only in an assigned territory. Such exclusive territories virtually eliminate so-called intra-brand competition, but this might be necessary in order to facilitate a more than compensating increase in inter-brand competition. In general then, competition agencies keep their eye on the ultimate objective and are prepared to apply a kind of cost-benefit analysis to arrangements which may appear somewhat anti-competitive. Competition agencies tend to apply rigid rules only where experience has shown that a certain practice has almost no potential to generate net advantages to consumers.

Lessons for the dispute settlement procedure. Although there is no international body dealing with competition disputes and therefore no internationally established principles, the settlement of competition disputes is commonly based on the notion that the regulator should be guided by an economic analysis of the issue and that decisions should be taken on a case by case basis according to their consequence on agents' welfare. Less consideration is given to such principles in the settlement of sanitary and technical disputes.

Regulatory measures likely to hinder imports are sometimes called technical or non-tariff barriers (Hillman, 1991; Roberts and DeRemer, 1997).[32] However, Baldwin (1970) has suggested that non-tariff barriers should be defined as policies which reduce potential world revenue. According to this definition, policies which in practice restrict trade flows would not be regarded as non-tariff barriers if their effect was to correct market inefficiencies and increase world revenue. Mahé (1997) proposes extending the definition to include non-merchant effects. He suggests that measures whose elimination would cause welfare losses in some countries that are greater than welfare gains in other countries should be classified as non-tariff barriers. This definition is in line with both economic theory and the idea of using cost benefit analysis to arbitrate disputes.

When international agreements call into question national regulations whose effect is also to reduce market inefficiencies, the welfare effects may be analytically ambiguous (Thilmany and Barrett, 1997). If a WTO judgement, for example, results in an obligation to import products that do not satisfy consumers' ethical, environmental or cultural concerns, anti-selection mechanisms could cause substantial welfare losses (Box 1). In practice, this could involve consumer boycotts or rejections which would affect demand for all the goods concerned, both imported and domestic. Estimating overall costs and benefits would involve quantifying the different variations in welfare,

---

32. Roberts and DeRemer (1997) give a particularly broad definition, classing the following measures as technical barriers to trade: (i) measures which protect plant and animal health against risks resulting from food additives, toxins, pesticides, disease and pathogens; (ii) food safety measures designed to protect human life and health from contaminants, residues, additives, natural toxins and biotechnology-related diseases; (iii) measures against commercial fraud, including measures to specify the composition of certain products or packaging methods or to regulate product appellations; (iv) measures concerning product quality characteristics (other than safety), such as size, freshness and taste; (v) measures designed to protect resources that do not belong to a single country.

raising awkward technical problems. Nonetheless, it is possible for welfare losses to be greater than welfare gains at a global level. It would be paradoxical if trade liberalisation, introduced by an international organisation in the framework of the settlement of disputes, were to result in more trade but less welfare. In such cases, Baldwin's criterion could serve as a basis for settling disputes (Mahé, 1997). Practical implementation could be based on a cost benefit analysis which would seem to be more in line with the maximisation of collective welfare than are rigid principles derived from uniform scientific standards.

In addition, it is often argued that *Codex* standards and texts express policy choices and that such policy choices could extend to national policies in such areas as the environment, consumer concerns, animal welfare, and societal values. If these values were considered from an economic point of view, the debate might lead to more convergence in the different point of views.[33] Genuine consumer aversion for certain imported products, for sanitary or cultural reasons, is reflected in a willingness to pay in order to avoid the products. Giving this willingness to pay greater importance in the settlement of disputes, by comparing it with the costs to other economic agents, would make it possible to take account of consumer preferences and to internalise hitherto diffuse fears about globalisation. This could also help to prevent its detractors from linking trade liberalisation with an obligation to consume products that do not correspond to consumer aspirations. In this way, economic analysis would make it possible to objectify consumer aspirations, even if it raises a number of practical difficulties.[34]

5.   Considerations for further analysis

The preceding discussion raises a number of trade concerns related to food safety and quality issues and provides some indication of the possible contributions that could be made through the application of economic analysis.

---

33.   Since it is based on a revelation of individual preferences, cost-benefit analysis can be seen as a tool for organising many different pieces of information and points of views in a consistent framework. In this respect, microeconomics can be seen as a useful negotiation language (see Henry, 1984).

34.   The technical difficulties in economic assessment should not be overstated. The methodologies described in Box 2 have raised similar difficulty in the evaluation of environmental costs and benefits, and agreements on procedures have progressed (see National Oceanic and Atmospheric Administration (NOAA) panel, *Federal Register* 58,10).

This section attempts to bring some of these issues together and suggests a few key areas where further research is needed.

5.1. *Acceptance of international standards*

The objective of the SPS Agreement is to improve human animal and plant health, establish a multilateral framework of rules and disciplines to guide the development, adoption and enforcement of SPS measures in order to minimise their negative trade effects and to further the use of harmonised standards, guidelines and recommendations developed by the relevant international organisations. The SPS Agreement requires measures to be based on science. Standards set up by the relevant international organisations and their subsidiary bodies, in particular Codex, OIE and IPPC, play an important role since, according to Article 3.1 of the SPS Agreement and in order to harmonise SPS measures on as wide a basis as possible, Members shall base their SPS measures on the international standards, guidelines or recommendations developed by these organisations. Thus, the adequacy of these international standards, guidelines and recommendations has become an issue, given the new role imposed on them in the context of international trade disputes.

A deeper understanding of the processes of international standardisation and how it impacts on trade could be achieved by examining: the ability of these international bodies to respond to the current needs of world agro-food trade (taking into account their internal decision making process); how domestic players such as regulators or standards bodies make use of such technical regulations (including conformity assessment); and how this contributes to market openness and competitiveness in a globalising economy. One contentious area is the interpretation of the precautionary principle, in particular as set out in Article 5.7 of the SPS Agreement, on the basis of which Members may adopt sanitary and phytosanitary measures on a provisional basis while seeking additional information, where scientific evidence is insufficient.

5.2. *Social standards and consumer preferences*

There is a body of opinion arguing that trade liberalisation poses a threat to social or environmental policy objectives. The view is that trade should be controlled and that harmonisation or co-ordination of standards are prerequisites to allowing freer trade (Hillman, 1997). Such views can lead to attempts to restrict market access where conditions in exporting countries do not meet the standards of the importing countries. For example, the desire to offset domestic costs associated with social and environmental legislation with import fees to create a more "level playing field" and encourage developing countries to improve social conditions. In addition, consumers are increasingly concerned about the methods used in agricultural and food production. They want food to be produced in ways that ensure sustainable use of the resource base, protect the environment and ethical concerns, such as animal welfare and genetically modified organisms. The 1997 EU green paper, *The General Principles of Food Law in the European Union,* specifically invites comments on what initiatives should be pursued in meeting the new aspirations of consumers.

The extent to which emerging social concerns and consumer preferences are influencing regulations is not well understood. A review of different national approaches to emerging social concerns and consumer preferences that would focus on how food safety and quality regulations in general (and NTBs in particular) are being implemented is needed (e.g. a comparative analysis of regulatory mechanisms governing agricultural biotechnology in Member countries). The current EU and US food policy reviews would be a useful place to start, as would the new food regulatory agencies in Canada and France. Of key interest would be the extent to which international standards are used in this process and the way in which trade considerations are taken into account. The study could encompass one or more aspects of "governance" such as policy co-ordination, coherence, and sequencing, institutional arrangements and the political economy (e.g. role of consumer groups and NGOs, of media, of information and consultation).

5.3.   *Administrative and compliance costs*

The role of governments in matters relating to food has been substantial, especially in designing and monitoring/enforcing minimum standards of safety, quality and labelling. In recent years, the role of governments in this area has been re-examined, particularly as harmonisation and equivalence (mutual recognition) become more important due to increasing trade, global market pressures, new technologies, information flow and public participation. As regulation increases, there is a danger that government Ministries or independent agencies responsible for regulatory policy may expand in size and complexity. Little work has been done to measure the cost of the creation, review, administration and enforcement of such regulations. Given that there are wide differences across OECD countries in the regulatory framework and administrative structures, it would be interesting to assess the relative costs to taxpayers and consumers.

Perhaps more importantly, it would be useful to examine the compliance costs to industry, which can have a significant impact on the viability and international competitiveness of firms (especially small and medium size operations) and can be passed on to consumers in the form of higher food costs. Food safety and quality standards can take a number of forms (e.g. performance, specification standards). Given that the costs of compliance associated with food standards generally exceed administrative costs by many orders of magnitude, the flexibility of economic agents should be maximised to the extent possible. In general, it is preferable to maximise the freedom of firms to choose the manner in which they meet the specified regulatory objectives, allowing firms to minimise compliance costs and promoting innovation in compliance technology. A comparison of compliance costs on firms and consumers across countries and sectors could shed light on the significance of variations in compliance costs. This study also indicated a lack of information on the trade impacts of food regulations and additional work in this area would provide useful information on the positive and negative impacts of regulatory regimes on market access.

5.4. *The role of private standards*

In the area of food safety and quality, private standards often supplement, and even exceed, the requirements laid down by public food safety regulations. There is a continuing debate on the extent to which governments should intervene in regulating quality. In order to determine the role of private standards, a clear distinction must be made between regulations providing a public benefit and those conferring private benefits — which should be paid for by beneficiaries. In most cases, health and safety regulations are considered to provide a public good while standards relating to certification (via certification and labelling) are often considered best left to the industry which has better knowledge of market requirements.

The development of private standards raises important issues regarding the future direction of food safety and quality regulation. Private standards represent a shift in responsibility from public agencies to private industry. This raises questions over the degree to which regulation is driven by private rather than public interest considerations. Secondly, the growth of private regulation on top of public regulation effectively increases the range of standards with which firms must comply. This could significantly increase the total regulatory burden on the food system and consumers. Private regulations also pose a danger of misuse by domestic firms to reduce market access to foreign competition. The growing incidence of private standards justifies an examination of the changing role of private industry in food safety and quality, including the potential trade implications (e.g. market access, harmonisation, transparency, competition).

ANNEX 1. SPS AND TBT AGREEMENTS AND WTO DISPUTE SETTLEMENT BODY

Sanitary and Phytosanitary Agreement (SPS)[35]

The Agreement on the Application of Sanitary and Phytosanitary Measures sets out the basic rules for food safety and animal and plant health standards. It allows countries to set their own standards. But it also says regulations must be based on science. They should be applied only to the extent necessary to protect human, animal or plant life or health. And they should not arbitrarily or unjustifiably discriminate between countries where identical or similar conditions prevail. Member countries are encouraged to use international standards, guidelines and recommendations where they exist. However, members may use measures which result in higher standards if there is scientific justification. They can also set higher standards based on appropriate assessment of risks so long as the approach is consistent, not arbitrary. The agreement still allows countries to use different standards and different methods of inspecting products.

Key features

All countries maintain measures to ensure that food is safe for consumers, and to prevent the spread of pests or diseases among animals and plants. These sanitary and phytosanitary measures can take many forms, such as requiring products to come from a disease-free area, inspection of products, specific treatment or processing of products, setting of allowable maximum levels of pesticide residues or permitted use of only certain additives in food. Sanitary (human and animal health) and phytosanitary (plant health) measures apply to domestically produced food or local animal and plant diseases, as well as to products coming from other countries.

---

35.   SPS text extracted from the WTO Web site:
        http://www.wto.org/wto/goods/sps.htm.

Protection or protectionism? Sanitary and phytosanitary measures, by their very nature, may result in restrictions on trade. All governments accept the fact that some limits to trade may be necessary to ensure food safety and animal and plant health protection. However, governments are sometimes pressured to go beyond what is needed for health protection and to use sanitary and phytosanitary restrictions to shield domestic producers from economic competition. Such pressure is likely to increase as other trade barriers are reduced as a result of the Uruguay Round agreements. A sanitary or phytosanitary restriction which is not actually required for health reasons can be a very effective protectionist device, and because of its technical complexity, a particularly deceptive and difficult barrier to challenge. The Agreement on Sanitary and Phytosanitary Measures (SPS) builds on previous GATT rules to restrict the use of unjustified sanitary and phytosanitary measures for the purpose of trade protection. The basic aim of the SPS Agreement is to maintain the sovereign right of any government to provide the level of health protection it deems appropriate, but to ensure that these sovereign rights are not misused for protectionist purposes and do not result in unnecessary barriers to international trade.

Justification of measures. The SPS Agreement, while permitting governments to maintain appropriate sanitary and phytosanitary protection, reduces possible arbitrariness of decisions and encourages consistent decision-making. It requires that sanitary and phytosanitary measures be applied for no other purpose than that of ensuring food safety and animal and plant health. In particular, the agreement clarifies which factors should be taken into account in the assessment of the risk involved. Measures to ensure food safety and to protect the health of animals and plants should be based as far as possible on the analysis and assessment of scientific data.

International standards. The SPS Agreement encourages governments to establish national SPS measures consistent with international standards, guidelines and recommendations. This process is often referred to as "harmonisation". The WTO itself does not and will not develop such standards. However, most of the WTO's member governments (132 at the date of drafting) participate in the development of these standards in other international bodies. The standards are developed by leading scientists in the field and governmental experts on health protection and are subject to international scrutiny and review.

International standards are often higher than the national requirements of many countries, including developed countries, but the SPS Agreement explicitly permits governments to choose not to use the international standards. However, if the national requirement goes beyond international standards, a country may be asked to provide scientific justification, demonstrating that the relevant international standard would not result in the level of health protection the country considered appropriate.

Adapting to conditions. Due to differences in climate, existing pests or diseases, or food safety conditions, it is not always appropriate to impose the same sanitary and phytosanitary requirements on food, animal or plant products coming from different countries. Therefore, sanitary and phytosanitary measures sometimes vary, depending on the country of origin of the food, animal or plant product concerned. This is taken into account in the SPS Agreement. Governments should also recognise disease-free areas which may not correspond to political boundaries, and appropriately adapt their requirements to products from these areas. The agreement, however, checks unjustified discrimination in the use of sanitary and phytosanitary measures, whether in favour of domestic producers or among foreign suppliers.

Alternative measures. An acceptable level of risk can often be achieved in alternative ways. Among the alternatives -- and on the assumption that they are technically and economically feasible and provide the same level of food safety or animal and plant health -- governments should select those which are not more trade restrictive than required to meet their health objective. Furthermore, if another country can show that the measures it applies provide the same level of health protection, these should be accepted as equivalent. This helps ensure that protection is maintained while providing the greatest quantity and variety of safe foodstuffs for consumers, the best availability of safe inputs for producers, and healthy economic competition.

Risk assessment. The SPS Agreement increases the transparency of sanitary and phytosanitary measures. Countries must establish SPS measures on the basis of an appropriate assessment of the risks involved, and, if requested, make known what factors they took into consideration and the assessment procedures they used. Although many governments already use risk assessment in their management of food safety and animal and plant health, the SPS Agreement encourages the wider use of systematic risk assessment among all WTO member governments and for all relevant products.

Transparency. Governments are required to notify other countries of any new or changed sanitary and phytosanitary requirements which affect trade, and to set up offices (called "Enquiry Points") to respond to requests for more information on new or existing measures. They also must open to scrutiny how they apply their food safety and animal and plant health regulations. The systematic communication of information and exchange of experiences among the WTO's member governments provides a better basis for national standards. Such increased transparency also protects the interests of consumers, as well as of trading partners, from hidden protectionism through unnecessary technical requirements.

A special Committee has been established within the WTO as a forum for the exchange of information among member governments on all aspects related to the implementation of the SPS Agreement. The SPS Committee reviews compliance with the agreement, discusses matters with potential trade impacts, and maintains close co-operation with the appropriate technical organisations. In a trade dispute regarding a sanitary or phytosanitary measure, the normal WTO dispute settlement procedures are used, and advice from appropriate scientific experts can be sought.

Agreement on Technical Barriers to Trade (TBT)[36]

The Agreement on Technical Barriers to Trade (TBT) was signed in 1979, but its scope was enormously increased by the Uruguay Round, not least because henceforth all WTO members must comply with it. Furthermore, countries cannot reject panel conclusions simply because they are unfavourable.

The TBT Agreement covers technical regulations, standards and conformity assessment procedures. Its scope extends to all traded goods and concerns all technical regulations and standards, including packaging, marking and labelling. In the agro-food sector, the TBT Agreement applies to all rules other than those relating to animal, plant and human life and health, which are covered by the SPS Agreement. The SPS Agreement constitutes an exception to the general scope of the TBT Agreement; everything not covered by the SPS Agreement continues to be covered by the TBT Agreement. The scope of the SPS Agreement is extremely precise, and issues such as nutrition or the allergenic properties of certain nutriments are covered by the TBT Agreement not the SPS Agreement. Furthermore, the TBT Agreement applies to everything that does not explicitly concern health (packaging, composition in relation to certain appellations, nutritional labelling, etc.).

The *TBT Committee*, originally made up of all the signatory States of the GATT Tokyo Round Agreements, has been in existence since 1980. Like the SPS Committee, it is now open (since 1995) to all the WTO members as the Agreement now applies to all signatories of the Uruguay Round Agreement. However, the two committees are open to governments only and not to non-governmental organisations. The TBT Committee is responsible for

---

36.   This section draws from Doussin (1995).

supervising implementation of the TBT Agreement. As regards regulatory measures, the Committee monitors compliance with the following principles:

- national regulations must not discriminate unjustifiably between products on account of their origin;

- measures must have a legitimate aim and achieve it in such a way as to minimise the restrictions on trade;

- favourable treatment is given to States which comply with the relevant international standards. Non-compliance may be legitimate, but in such cases there is a transparency obligation and other States must be notified of the proposed regulations so that they can comment on them. The State in question must establish that the desired aim is legitimate and that the proposed measures are appropriate (see preceding point).

The stance of the TBT Agreement with regard to conformity assessment is the same as for regulations: unjustified discrimination (e.g. deadlines for and cost of admission and inspection procedures, excessive controls) is prohibited, and information must be provided (complaint review procedure, onus of transparency and proof on the country departing from international recommendations).

In practice, the TBT Committee is the centralising body for notifications when a country introduces regulations that depart from international standards and for comments from other States following such notifications. The TBT Committee also serves to facilitate negotiations between States. Most of the problems that arise can be settled at its biannual meetings; the use of panels to settle disputes remains the exception rather than the rule.

Dispute Settlement Body (DSB)

Disputes should first be discussed, and if possible resolved, bilaterally. However, the WTO may act as a dispute settlement body. The DSB is composed of all WTO members and meets almost monthly. It is informed about consultations in progress and hears applications for formal dispute settlement. The DSB may be asked to set up a panel (*i.e.* a group of experts) to consider the dispute. This procedure may only be initiated by member States. The panel delivers a report to the WTO and to the two parties, who may submit comments. The WTO adopts the report unless there is a unanimous decision to reject it. Provision has been made for an appeal procedure. The WTO monitors implementation of the report's recommendations, which indicate the measures that must be brought into conformity with WTO rules, may include compensation or, failing that, retaliatory measures. If measures are not brought into conformity within a reasonable period of time, the DSB can agree to members negotiating temporary measures of compensation or, failing that, retaliatory measures (again temporary) being introduced. In such cases, all issues would be kept under constant surveillance by all members through the DSB. The members of the panel are trade or legal experts but may call on "technical" experts. The creation of the DSB in 1995 should lead to the formation of a body of case law and help to clarify the occasionally imprecise terms of the SPS and TBT Agreements.

ANNEX 2. *CODEX ALIMENTARIUS*[37]

The *Codex Alimentarius* was created by the FAO (Food and Agriculture Organization) and the WHO (the World Health Organisation) in 1962 and is financed by them. The *Codex Alimentarius* itself is a set of "norms". *Codex* is generally understood to mean the *Codex Alimentarius* Commission or CAC. It consists of:

- A Commission composed of 165 countries, who take final decisions on the adoption of texts. This Commission meets every two years, in Rome and Geneva alternately.

- An Executive Committee that has 10 members, representing the different geographical areas of the world and which does the preparatory work. It meets every year and guides the work of Codex.

- A permanent Secretariat of six officials that is based in Rome. It follows the work of the different subsidiary bodies.

- Committees which meet to discuss norms and other texts. Some of these committees are horizontal in nature, dealing with categories of regulations (additives, labelling, etc.) or general principles such as inspection procedures or analytical methods. Others are concerned with products or product groups (dairy products or seafood, etc.) Five committees are involved in problems specific to the different geographical zones (Africa, Latin America and the Caribbean, North America and the Pacific, Asia, and Europe). Each committee is chaired and hosted by an individual country, which organises the meetings.

---

37. This annex extracts a number of elements from J.P. Doussin (1995) and from FAO Web sites: http://www.fao.org/org/news/1999/codex-e.htm, and http://www.fao.org/doocrep/w9114e/W9114e02.htm#TopOfPage.

The Commission of *Codex Alimentarius* establishes norms, directives, recommendations, or codes of practice on which countries can agree with a view to "protecting the health of consumers and to ensure that procedures followed in trade of food products are fair". The texts drawn up under the auspices of *Codex* are intended to "guide and promote the development, implementation and harmonisation of definitions and requirements covering food products with a view to facilitating international trade".

One of the principal purposes of *Codex* is the preparation of food standards. *Codex* adopts international recommended standards, guidelines and codes of practice after thorough consideration by all *Codex* member countries. The *Codex Alimentarius* contains more than 200 standards. There are general standards or recommendations for: food labelling; food additives; contaminants; methods of analysis and sampling; food hygiene; nutrition and foods for special dietary uses; food import and export inspection and certification systems; residues of veterinary drugs in foods; and pesticide residues in foods. *Codex* standards include:

- Food Standards for Commodities (237)

- Codes of Hygienic or Technological Practice (41)

- Pesticides Evaluated (185)

- Limits for Pesticide Residues (3 274)

- Guidelines for Contaminants (25)

- Food Additives Evaluated (1 005)

- Veterinary Drugs Evaluated (54)

Norms are developed according to an 8 stage process, the final stage being adoption by the Commission. Each country can simply adopt the norms in their entirety, adopt them with specified derogations or accept the free circulation of goods conforming with the norm without necessarily changing their own domestic regulations. The final phase involves adoption and gives the status of a norm to the measure in question. Up to that point, it is only a collection of proofs and scientific recommendations that has no particular legal status in the context of the SPS Agreement, but which facilitates countries in resolving disputes or controversies as they arise.

The *Codex* Commission also advises governments with a view to protecting the health of consumers and harmonising regulations in order to facilitate trade. Through a specific committee the *Codex* Commission is also involved in inspection services and in the certification of imported foodstuffs.

It develops directives and recommendations concerning the transparency of procedures, evaluating different methods with a view to establishing equivalence and mutual recognition between different national systems, harmonisation of checks and controls, of measures taken and of official certifications.

An increasing number of countries are aligning their national food standards, or parts of them with those of the Codex. This is particularly so in the case of additives, contaminants and residues. The World Trade Organization (WTO) Agreements on the Application of Sanitary and Phytosanitary Measures (SPS) and on Technical Barriers to Trade (TBT) encourage the international harmonization of food standards on the basis of *Codex* standards. The work of the *Codex* Commission goes beyond means of removing trade barriers. It also encourages countries to adopt ethical practices. The Code of Ethics for International Trade in Food, for example, calls on parties to stop dumping poor-quality or unsafe food on to international markets.

Many countries need FAO/WHO advice and recommendations about the risks that can be caused by chemicals that become, intentionally or unintentionally, part of foods. The Joint FAO/WHO Expert Committee on Food Additives (JECFA) advises the *Codex* Commission on food additives, contaminants and residues of veterinary drugs. JECFA establishes the amount of an additive that can be ingested on a daily basis, even for a lifetime, without significant risk. JECFA is independent of the Commission. It has examined more than 700 chemicals as well as 25 contaminants. JECFA members are selected from the scientific community. They must be impartial and work as individuals and not as representatives of their governments or institutions. Another group of scientists (Joint FAO/WHO Meeting on Pesticide Residues) advises the Commission on pesticide residues.

Codex activities of the future will differ considerably from what they have been until now. Scientific developments in fields relating to food, changing attitudes of consumers, new approaches to food control, changing perceptions of government and food industry responsibilities and changing food quality and safety concepts will present the Commission with new challenges and, conceivably, the need for new standards. The consumer protection elements of the *Codex Alimentarius*, which are domain of the "horizontal" committees, are currently gaining in importance. The application of biotechnology to food processing and production of raw food materials is already under scrutiny by the Commission, which is continually examining new concepts and systems associated with food safety and the protection of consumers against health hazards. These topical matters provide some insight into the direction that the Commission's activities are likely to take in the future.

ANNEX 3. THE INTERNATIONAL PLANT PROTECTION CONVENTION (IPPC)[38]

The IPPC was set up in 1952 to facilitate the adoption of international sanitary and phyto-sanitary standards. Currently 110 countries are members. Its objective is to ensure "effective common action to prevent the spread and introduction pests and parasites affecting plants and plant products and to promote control measures".

The Convention is an international treaty which obliges its signatories to take measures to ensure the safety of imports and exports of plants or plant products likely to contain pests or diseases. In particular countries are committed to set up the appropriate institutions, put the necessary checks, certification and disinfection procedures in place and to make information about them widely available. However, IPPC is often taken to mean the secretariat of the IPPC. It is based at the FAO in Rome in the plant protection service of the Division for the Production and Protection of Vegetable Products. Among its roles, facilitating trade and preventing the use of unjustified trade barriers figure specifically. The responsibilities of the IPPC Secretariat are as follows:

- to reinforce international co-operation with respect to the Convention;

- to develop international standards for phytosanitary measures;

- the centralisation and dissemination of information on plant parasites that could be contained in imports (pest quarantine);

- to provide technical assistance to developing countries.

Norms are adopted by the FAO Conference, which brings together all the member countries of the FAO. Norms established in this way are

---

38.   This annex extracts a number of elements from the FAO Web site: http://www.fao.org/ag/agp/agpp/pq.

recognised by the SPS Agreement. They are subject to periodical revision by the expert committee on phytosanitary measures of the FAO.

The IPPC was amended once in 1979 and again in 1997. The revision of the IPPC approved in 1997 reflects an updating of the Convention to reflect contemporary phytosanitary concepts and the role of the IPPC in relation to the Uruguay Round Agreements of the World Trade Organization, particularly the Agreement on the Application of Sanitary and Phytosanitary Measures (the SPS Agreement). The SPS Agreement identifies the IPPC as the organization providing international standards for measures implemented by governments to protect their plant resources from harmful pests (phytosanitary measures). The IPPC complements the SPS Agreement by providing the international standards that help to ensure that phytosanitary measures have a scientific basis for their placement and strength and are not used as unjustified barriers to international trade.

The New Revised Text of the IPPC (NRT) emphasizes co-operation and the exchange of information toward the objective of global harmonization. With the revision of the IPPC in 1997, information exchange on plant pests (defined by the New Revised Text of the IPPC as "any species, strain or biotype of plant, animal or pathogenic agent injurious to plants or plant products") is entering a new era, with the IPPC Secretariat providing the forum for the sharing of information by Member countries. In the past, it had been the responsibility of the Member countries to inform the FAO/IPPC Secretariat of their phytosanitary regulations, any changes to phytosanitary regulations and pest status, who acted as the centralised depository for this information. This system was not efficient in information dissemination and many countries did not meet their obligations. The WTO-SPS Agreement and the Revised IPPC now ensures that there are structured channels for notification of changes to phytosanitary measures, deviations in the relevant regulations, and a forum for information sharing. The primary responsibility on information sharing now lies with each individual country Member.

In addition to describing national plant protection responsibilities it also addresses important elements of international co-operation for the protection of plant health and the establishment and use of International Standards for Phytosanitary Measures (ISPMs), including:

- Principles of Plant Quarantine as Related to International Trade

- Guidelines for Pest Risk Analysis

- Code of Conduct for the Import and Release of Exotic Biological Control Agents

- Requirements for the Establishment of Pest Free Areas
- Glossary of Phytosanitary Terms
- Guidelines for Surveillance
- Export Certification System
- Determination of Pest Status in an Area
- Guidelines for Pest Eradication Programmes

Although the IPPC has strong implications for international trade, it has international co-operation for plant protection as its focus. Many forms of co-operation fall within the scope of the Convention. Its application to plants is not limited only to the protection of cultivated plants or direct damage from pests. The scope of the Convention extends to the protection of cultivated and natural flora as well as plant products, and includes both direct and indirect damage by pests.

The IPPC has always played an important role international trade. The Convention has encouraged countries to ensure through phytosanitary certification that their exports are not the means for introducing new pests to their trading partners. Likewise, importing countries strive to ensure that measures they have in place for protection are technically justified.

The relationship of the IPPC to international trade is strengthened by the WTO-SPS Agreement because it names the IPPC as the international organization responsible for phytosanitary standard-setting and the harmonization of phytosanitary measures affecting trade. Both agreements are distinct in their scope, purpose, and membership. Neither agreement is supplementary to the other. Instead, they are complementary in the areas where they overlap. The SPS Agreement makes provision for plant protection in a trade agreement, and the IPPC makes provision for trade in a protection agreement.

The Convention is a legally binding agreement, but standards developed and adopted by the IPPC are not legally binding. However, measures that are based on international standards do not require supporting justification. Measures that deviate from international standards, or measures that exist in the absence of international standards must be based on scientific principles and evidence. Emergency (or provisional) measures may be taken without such analyses, but must be reviewed for their scientific justification and modified accordingly to be legitimately maintained.

The IPPC also has dispute settlement provisions in the instance where measures may be challenged as unjustified barriers to trade. The dispute settlement process under the IPPC offers an alternative for examining controversial issues at a technical level. Although the dispute settlement process in the IPPC is non-binding, the results of the process can be expected to have substantial influence in disputes that may be raised to the WTO level under the SPS Agreement.

ANNEX 4. THE INTERNATIONAL OFFICE OF EPIZOOTICS (OIE)

The IOE was created in 1924 to facilitate trade in animals and animal products, both with a view to protecting the health of consumers and preventing the spread of diseases.   This intergovernmental organisation currently has 147 member countries. It is based in Paris. Unlike the IPPC, the OIE is not part of the United Nations. It is financed by contributions from its member countries. Its role is:

- to inform member governments about the occurrence and course of animal diseases in the world and how to combat them;

- at international level to co-ordinate studies on the monitoring and control of diseases;

- to study regulations governing trade in animals and animal products with a view to harmonisation among countries.

The highest organ of the OIE is the International Committee, composed of delegates from the member countries. It meets in general session once a year. Each member has one vote. The Central Bureau applies resolutions developed by the Committee with the assistance of three specialised, elected commissions (Commission on Foot and Mouth Disease and other Epizootic Diseases, Commission on Standards, Commission on Fish Diseases). Another Commission -- the International Animal Health Code is more involved with regulation. Four working groups (biotechnology, computerisation and epidemiology, the registration of veterinary medicines, wildlife diseases) contribute to the dissemination of information in member countries. Regional offices help to control diseases. A certain number of laboratories across the world are certified as reference laboratories and collaborating centres, supplying scientific and technical support on monitoring and control of animal diseases to governments.

The OIE is above all an observatory of animal health. To provide information to national veterinary services is its priority mission. It centralises information on the basis of two lists of diseases, according to how serious they are. In the case of epizootic diseases, the country in which an outbreak has occurred must inform the OIE within 24 hours. The OIE published weekly information on epidemiological and sanitary incidents. The OIE establishes and maintains the lists of countries recognised as being free of certain of the most serious diseases, notably foot and mouth disease.

The OIE establishes norms whose use is explicitly recommended in the SPS Agreement. Specifically the OIE publishes the International Animal Health Code adopted by the Committee which defines sanitary norms for international trade in animals and animal products. The OIE Code establishes the conditions whereby a region may be declared free of a certain disease. National veterinary administrations must inform the OIE of the provisions of the quarantine and sanitary regulations governing imports and exports. The OIE assists in the harmonisation of regulations applicable to trade in animals and animal products. In addition to the International Animal Health Code, the OIE publishes a manual, developed by the Commission on Standards, presenting standardised methods for diagnostic tests and vaccines to be applied in international trade. It is the recognised international reference on this subject.

ANNEX 5.   AGREEMENT ON TRADE-RELATED ASPECTS OF INTELLECTUAL PROPERTY RIGHTS (TRIPs)[39]

What are intellectual property rights?

Intellectual property rights are the rights given to persons over the creations of their minds. They usually give the creator an exclusive right over the use of his/her creation for a certain period of time.

Intellectual property rights are customarily divided into two main areas:

i. **Copyright and rights related to copyright**. The rights of authors of literary and artistic works (such as books and other writings, musical compositions, paintings, sculpture, computer programs and films) are protected by copyright, for a minimum period of 50 years after the death of the author. Also protected through copyright and related (sometimes referred to as "neighbouring") rights are the rights of performers (e.g. actors, singers and musicians), producers of phonograms (sound recordings) and broadcasting organizations. The main social purpose of protection of copyright and related rights is to encourage and reward creative work.

ii. **Industrial property**. Industrial property can usefully be divided into two main areas:

One area can be characterized as the protection of distinctive signs, in particular trademarks (which distinguish the goods or services of one undertaking from those of other undertakings) and geographical indications (which identify a good as originating in a place where a given characteristic of the good is essentially

---

39.   Text extracted from the WTO Web site:
        http://www.wto.org/wto/intellec/intellec.htm.

attributable to its geographical origin). The protection of such distinctive signs aims to stimulate and ensure fair competition and to protect consumers, by enabling them to make informed choices between various goods and services. The protection may last indefinitely, provided the sign in question continues to be distinctive.

Other types of industrial property are protected primarily to stimulate innovation, design and the creation of technology. In this category fall inventions (protected by patents), industrial designs and trade secrets. The social purpose is to provide protection for the results of investment in the development of new technology, thus giving the incentive and means to finance research and development activities. A functioning intellectual property regime should also facilitate the transfer of technology in the form of foreign direct investment, joint ventures and licensing. The protection is usually given for a finite term (typically 20 years in the case of patents).

While the basic social objectives of intellectual property protection are as outlined above, it should also be noted that the exclusive rights given are generally subject to a number of limitations and exceptions, aimed at fine-tuning the balance that has to be found between the legitimate interests of right holders and of users.

Main features of the TRIPs Agreement

The TRIPs Agreement, which came into effect on 1 January 1995, is to date the most comprehensive multilateral agreement on intellectual property. The areas of intellectual property that it covers are: copyright and related rights (*i.e.* the rights of performers, producers of sound recordings and broadcasting organizations); trademarks including service marks; geographical indications including appellations of origin; industrial designs; patents including the protection of new varieties of plants; the layout-designs of integrated circuits; and undisclosed information including trade secrets and test data.

The three main features of the Agreement are:

i. **Standards**. In respect of each of the main areas of intellectual property covered by the TRIPs Agreement, the Agreement sets out the minimum standards of protection to be provided by each Member. Each of the main elements of protection is defined, namely the subject-matter to be protected, the rights to be conferred and permissible exceptions to those rights, and the minimum duration of protection. The Agreement sets these standards by requiring, first, that the substantive obligations of the main conventions of the WIPO, the Paris Convention for the Protection of Industrial Property (Paris Convention) and the Berne Convention for the Protection of Literary and Artistic Works (Berne Convention) in their most recent versions, must be complied with. With the exception of the provisions of the Berne Convention on moral rights, all the main substantive provisions of these conventions are incorporated by reference and thus become obligations under the TRIPs Agreement between TRIPs Member countries. The relevant provisions are to be found in Articles 2.1 and 9.1 of the TRIPs Agreement, which relate, respectively, to the Paris Convention and to the Berne Convention. Secondly, the TRIPs Agreement adds a substantial number of additional obligations on matters where the pre-existing conventions are silent or were seen as being inadequate. The TRIPs Agreement is thus sometimes referred to as a Berne and Paris-plus agreement.

ii. **Enforcement**. The second main set of provisions deals with domestic procedures and remedies for the enforcement of intellectual property rights. The Agreement lays down certain general principles applicable to all IPR enforcement procedures. In addition, it contains provisions on civil and administrative procedures and remedies, provisional measures, special requirements related to border measures and criminal procedures, which specify, in a certain amount of detail, the procedures and remedies that must be available so that right holders can effectively enforce their rights.

iii. **Dispute settlement**. The Agreement makes disputes between WTO Members about the respect of the TRIPs obligations subject to the WTO's dispute settlement procedures.

In addition the Agreement provides for certain basic principles, such as national and most-favoured-nation treatment, and some general rules to ensure that procedural difficulties in acquiring or maintaining IPRs do not nullify the substantive benefits that should flow from the Agreement. The obligations under the Agreement will apply equally to all Member countries, but developing countries will have a longer period to phase them in. Special transition arrangements operate in the situation where a developing country does not presently provide product patent protection in the area of pharmaceuticals.

The TRIPs Agreement is a minimum standards agreement, which allows Members to provide more extensive protection of intellectual property if they so wish. Members are left free to determine the appropriate method of implementing the provisions of the Agreement within their own legal system and practice.

Certain general provisions

As in the main pre-existing intellectual property conventions, the basic obligation on each Member country is to accord the treatment in regard to the protection of intellectual property provided for under the Agreement to the persons of other Members. Article 1.3 defines who these persons are. These persons are referred to as "nationals" but include persons, natural or legal, who have a close attachment to other Members without necessarily being nationals. The criteria for determining which persons must thus benefit from the treatment provided for under the Agreement are those laid down for this purpose in the main pre-existing intellectual property conventions of WIPO, applied of course with respect to all WTO Members whether or not they are party to those conventions. These conventions are the Paris Convention, the Berne Convention, International Convention for the Protection of Performers, Producers of Phonograms and Broadcasting Organizations (Rome Convention), and the Treaty on Intellectual Property in Respect of Integrated Circuits (IPIC Treaty).

Articles 3, 4 and 5 include the fundamental rules on national and most-favoured-nation treatment of foreign nationals, which are common to all categories of intellectual property covered by the Agreement. These obligations cover not only the substantive standards of protection but also matters affecting the availability, acquisition, scope, maintenance and enforcement of intellectual property rights as well as those matters affecting the use of intellectual property rights specifically addressed in the Agreement. While the national treatment clause forbids discrimination between a Member's own nationals and the nationals of other Members, the most-favoured-nation treatment clause forbids discrimination between the nationals of other Members. In respect of the national treatment obligation, the exceptions allowed under the pre-existing intellectual property conventions of WIPO are also allowed under TRIPs. Where these exceptions allow material reciprocity, a consequential exception to MFN treatment is also permitted (e.g. comparison of terms for copyright protection in excess of the minimum term required by the TRIPs Agreement as provided under Article 7(8) of the Berne Convention as incorporated into the TRIPs Agreement). Certain other limited exceptions to the MFN obligation are also provided for.

The general goals of the TRIPs Agreement are contained in the Preamble of the Agreement, which reproduces the basic Uruguay Round negotiating objectives established in the TRIPs area by the 1986 Punta del Este Declaration and the 1988/89 Mid-Term Review. These objectives include the reduction of distortions and impediments to international trade, promotion of effective and adequate protection of intellectual property rights, and ensuring that measures and procedures to enforce intellectual property rights do not themselves become barriers to legitimate trade. These objectives should be read in conjunction with Article 7, entitled "Objectives", according to which the protection and enforcement of intellectual property rights should contribute to the promotion of technological innovation and to the transfer and dissemination of technology, to the mutual advantage of producers and users of technological knowledge and in a manner conducive to social and economic welfare, and to a balance of rights and obligations. Article 8, entitled "Principles", recognizes the rights of Members to adopt measures for public health and other public interest reasons and to prevent the abuse of intellectual property rights, provided that such measures are consistent with the provisions of the TRIPs Agreement.

BIBLIOGRAPHY

AKERLOF, G. (1970), "The Market for Lemons: Qualitative Uncertainty and the Market Mechanism", *Quarterly Journal of Economics*, 84,1 : 488-500.

ANTLE, J.M. (1995), *Choice and Efficiency in Food Safety Policy,* The AEI Press American Enterprise Institute, Washington D.C.

ARROW, K.J., M.L. CROPPER, G.C. EADS, R.W. HAHN, L.B. LAVE, R.G. NOLL, P.R. PORTNEY, M. RUSSELL, R. SCHMALENSEE, V.K. SMITH et R.N. STAVINS (1996), "Is there a Role for Benefit-Cost Analysis in Environmental, Health and Safety Regulation" ?, *Science*, 272 : 221-222.

BAGHWATI, J.N. and R.E. HUDEC (1997), *Fair Trade and Harmonization, Prerequisites for Free Trade* ?, MIT Press, Cambridge, Mass.

BALDWIN, R.E. (1970), *Non-Tariff Distortions in International Trade,* Brookings Institute, Washington.

BEALES, H., R. CRASWELL and S. SALOP (1981), "The Efficient Regulation of Consumer Information", *Journal of Law and Economics*, vol. XXIV, December, pp 491-544.

BUREAU, J.C., S. MARETTE and A. SCHIAVINA. (1998), "Non-Tariff Trade Barriers and Consumers' Information: the Case of EU-US Trade Dispute on Beef", forthcoming in *European Review of Agricultural Economics.*

BUZBY, J.C., T. ROBERTS, C.J. LIN and J. MACDONALD (1996), "Bacterial Foodborne Disease: Medical Costs and Productivity Losses", U.S. Department of Agriculture, Economic Research Service, *Agricultural Economic Report* 741.

CASWELL, J.A. and N.H. HOOKER (1996), "HACCP as an International Trade Standard", *American Journal of Agricultural Economics*, 78, August, pp 775-79.

CASWELL, J.A. and E.M. MOJDUSKA (1996), Using Informational Labelling to Influence the Market for Quality in Food Products, AAEA meeting, July 29, San Antonio, forthcoming in *American Journal of Agricultural Economics*.

CHEN, J. (1996), "A Sober Second Look at Appelations of Origin: How the United States will Crash France's Wine and Cheese Party", *Minnesota Journal of Global Trade*, Winter, Vol. 5,1. (French version : "Le statut légal des appellations d'origine contrôlées aux États-Unis d'Amérique", *Revue de droit rural*, 249, January 1997, pp. 35-43.

CODEX ALIMENTARIUS COMMISSION (1999), *Guidelines for the Production, Processing, Labelling and Marketing of Organically Produced Foods*, CAC/GL 32-1999, FAO/WHO Foods Standards Programme, Rome.

COMMISSION EUROPEENNE (1997), Report on the United States Barriers to Trade and Investment 1997, Directorate General for External Relations, European Commission, Bruxelles, July, 1997.

CRUTCHFIELD, S., J.C. BUZBY, T. ROBERTS, M. OLLINGER and LIN, C.T.J. (1997), "An Economic Assessment of Food Safety Regulations: the New Approach to Meat and Poultry Inspection", U.S. Department of Agriculture, Economic Research Service, *Agricultural Economic Report* 755.

DOUSSIN, J.P. (1995), "Le *Codex Alimentarius* à l'heure de l'Organisation mondiale du commerce", *Annales des Falsifications de l'Expertise Chimique et Toxicologique*, 933, pp 281-292.

DOUSSIN, J.P. (1998),. "Réglementations et institutions dans les aspects sanitaires et phytosanitaires du commerce international". Seminar on "Normes et obstacles techniques aux échanges", Institut National de la Recherche Agronomique, Grignon, 27-28 Janvier 1998.

FALVEY, R.E. (1989), "Trade, Quality Reputations and Commercial Policy", *International Economic Review*, 30,3, August, pp 607-622.

FREEMAN, A.M. (1993), *Measuring Environmental and Resource Values. Theory and Methods,* Resources for the Future, Washington D.C.

GAL-OR, E. (1989a), "Quality and Quantity Competition", *The Bell Journal of Economics*, pp 590-600.

GAL-OR, E. (1989b), "Warranties as a Signal of Quality", *Canadian Journal of Economics*, février, 22,1, pp 50-61.

GAO (1997), *Agricultural Exports. US Needs a More Integrated Approach to Address Sanitary/phytosanitary Issues*, Report to Congressional Requesters 98-32, United States General Accounting Office, Washington D.C.

GODARD, O. ED. (1997), *Le principe de précaution dans la conduite des affaires humaines,* Éditions de la Maison des Sciences de l'Homme, Paris.

GRABOWSKI, H.G. and J.M. VERNON (1983), *The Regulation of Pharmaceuticals: Balancing the Benefits and Risks,* American Enterprise Institute, Washington D.C.

GROSSMAN, G. and C. SHAPIRO (1988), "Counterfeit Product Trade", *American Economic Review*, 75, pp 59-76.

HARRINGTON, W. and P.R. PORTNEY (1987), "Valuing the Benefits of Health and Safety Regulations", *Journal of Urban Economics*, 22, 1, pp 101-112.

HAYES, D., J. SHORGREN, S. SHIN and J. KLIEBENSTEIN (1995), "Valuing Food Safety in Experimental Auctions Markets", *American Journal of Agricultural Economics*, 77, 1, February, pp 40-53.

HENRY, C. (1991), *Microeconomics for Public Policy: Helping the Invisible Hand,* Clarendon Press, Oxford.

HENRY, C. (1984), "La microéconomie comme langage et enjeu de négociations" *Revue économique*, 35, pp 177-187.

HENSON, S. (1998). "Regulating the Trade Effects of National Food Safety Standards: Discussion of some Issues", OCDE Workshop on Emerging Trade Issues in Agriculture, Paris, October 25-27.

HILLMAN, J. (1978), *Nontariff Agricultural Trade Barriers,* University of Nebraska Press, Lincoln.

HILLMAN, J. (1991), *Technical Barriers to Agricultural Trade,* Westview Press, Boulder.

HILLMAN, J. (1997), "Nontariff Agricultural Trade Barriers Revisited", in *Understanding Technical Barriers to Agricultural Trade*, Orden, D. and Roberts, D. eds., The International Agricultural Trade Research Consortium, University of Minnesota, St Paul.

HOBAN, T.J. (1997), "Consumer Acceptance of Biotechnology: An International Perspective", *Nature Biotechnology*, 15, mars 1997, pp. 232-235.

HOOKER, N.H. and J.A. CASWELL (1996), "Voluntary and Mandatory Quality Management Systems in Food Processing", Working Paper, U-Mass, Ahmerst.

HUDEC, R.E. (1997), "Does the Agreement on Agriculture Work? Agricultural Disputes after the Uruguay Round", International Agricultural Trade Research Consortium, San Diego, Dec. 14-16.

JAMES, S. and K. ANDERSON (1998). "On the Need for more Economic Assessment of Quarantine/SPS Policies", CIES Seminar Paper 98-02, University of Adelaide.

JOHNSON, R.W.M. (1997), "Technical Measures for Meat and Other Products in Pacific Basin Countries", *Understanding Technical Barriers to Agricultural Trade*, D. Orden and D. Roberts, eds, The Internatioanal Agricutlural Trade Research Consortium, St Paul, Minnesota.

KOPP, R.J., A.J. KRUPNICK and M. TOMAN (1997), "Cost Benefit Analysis and Regulatory Reform: An Assessment of the Science and the Art", Discussion paper 97-19, Resources for the Future, Washington D.C.

KRAMER, C. (1989), "Food Safety and International Trade: the US-EC Meat and Hormone Controversies", in *The Political Economy of US Agriculture*, C.S. Kramer ed., Resources for the Future, Washington D.C.

KROUSE, C.G. (1990), *Theory of Industrial Economics,* Basil Blackwell, Cambridge, Mass.

KRUGMAN, P. (1979), "Increasing Returns, Monopolistic Competition and International Trade", *Journal of International Economics*, 9, pp 469-79.

LANDELFELD, J.S. and E.P. SESKIN (1982), "The Economic Value of Life: Linking Theory and Practice", *American Journal of Public Health*, 6, pp 555-66.

LACARTE-MURO, J. (1998). EC Measures concerning meat and meat products (hormones), Award of the Arbitrator. WT/DS26/15, WT/DS48/13, World Trade Organization, Geneva.

LORVELLEC, L. (1997), "Back to the Fields after the Storm: Agriculture in the European Union after the Uruguay Round Agreement", *Drake Journal of Agricultural Law*, 2, 2, Winter, pp 411-429.

LUTZ, N.A. (1989), "Warranties as Signals under Consumer Moral Hazard", *Rand Journal of Economics*, Summer, 20,2, pp 239-54.

MAGAT, W.A. and W.K. VISCUSI (1993), *Informational Approaches to Regulation,* Cambridge, MIT Press.

MAGAT, W.A., A.J. KRUPNICK and W.W. HARRINGTON (1986), *Rules in the Making: A Statistical Analysis of Regulatory Agency Behavior,* Resources for the Future, Washington D.C.

MAHE, L.P. (1997), "Environment and Quality Standards in the WTO. New Protectionism in Agricultural Trade", *European Review of Agricultural Economics*, 24, 3-4, pp 480-503.

MAHE, L.P. and F. ORTALO-MAGNE (1998). "International Co-operation in the Regulation of Food Quality and Safety AQttributes", OCDE Workshop on Emerging Trade Issues in Agriculture, Paris, October 25-27.

MARGOLIS, H. (1996). *Dealing with Risk,* University of Chicago Press, 1996.

MAZE, A., M.T. LETABLIER and E. VALCESCHINI (1996), "Les bases institutionnelles", dans Casabiance, F and Valceschini, E., *La qualité dans l'agro--alimentaire : émergence d'un champ de recherches,* INRA, Département systèmes agraires et développement, Paris.

MAZUREK, J.V. (1996), "The Role of Health Risk Assessment and Cost-Benefit Analysis in Environmental Decision Making in Selected Countries: An Initial Survey", Discussion paper 96-36. Resources for the Future, Washington D.C.

NDAYISENGA, F. and J. KINSEY (1994), "The Structure of Non Tariff Trade Measures on Agricultural Products in High-Income Countries", *Agribusiness*, 10, pp. 275-92.

OECD (1997), "Uses of Food Labelling Regulations", Paris.

OECD (1997a), "Regulatory Reform in the Agro-Food Sector", *Regulatory Reform Volume I: Sectoral Studies*, pp. 233-274, Paris.

OECD (1997b), "The Costs and Benefits of Food Safety Regulations: Fresh Meat Hygiene Standards in the United Kingdom", Paris.

OECD (1998), "Agro-Food Trade and Regulatory Reform An Overview", *Regulatory Reform in the Global Economy: Asian and Latin Perspectives*, pp. 50-59, Paris.

OECD (1998a), "The Agricultural Outlook 1998-2003", Paris.

ORDEN, D. and D. ROBERTS, eds (1997), *Understanding Technical Barriers to Agricultural Trade*, Proceedings of the International Agricultural Trade Research Consortium, University of Minnesota, St Paul.

ORDEN, D. and E. ROMANO (1996), "The Avocado Dispute and Other Technical Barriers to Agricultural Trade under NAFTA", conference on NAFTA and Agriculture, San Antonio, Texas, November.

POLLAK, R.A. (1995). "Regulating risks", *Journal of Economic Literature*, March, 33, 1, pp 179-91.

POLLAK, R.A. (1998). "Imagined Risks and Cost-Benefit Analysis", *American Economic Review*, Papers and Proceedings, 88, 2, pp 376-79.

POWELL, M. (1997), "Science in Sanitary and Phytosanitary Dispute Resolution", Discussion paper 97-50, Resources for the future, Washington D.C.

RAY, P.H. (1996*), The Integral Culture Survey: A Study of the Emergence of Transformation Values in America*, Institute of Noetic Sciences, and Fetzer Institute.

REGE, V. (1994), "Gatt Law and Environment: Related Issues Affecting the Trade of Developping Countries", *Journal of World Trade*, 28, 3, pp 95-169.

REITZES, J.D. (1992), "Quality Choice, Trade Policy and Firm Incentives", *International Economic Review*, 33, 4, November, pp 817-35.

ROBERTS, D. (1997), "Implementation of the WTO Agreement on the Application of Sanitary and Phytosanitary Measures", International Agricultural Trade Research Consortium meeting, San Diego, 14-16 December.

ROBERTS, D. and K. DEREMER (1997), "Overview of Foreign Technical Barriers to US Agricultural Exports", Commercial Agriculture Division, Staff paper AGES-9705, Economic Research Service, US Department of Agriculture, Washington D.C.

ROBERTS, D. and D. ORDEN (1997), "Determinants of Technical Barriers to Trade: the Case of US Phytosanitary Restrictions on Mexican Avocados", 1972-1995, in *Understanding Technical Barriers to Agricultural Trade*, Orden, D. and Roberts, D. eds., The International Agricultural Trade Research Consortium, University of Minnesota, St Paul.

SALTER, L. (1988), *Mandated Science: Science and Scientists in the Making of Standards,* Dordecht, Kluwer.

SHAPIRO, C. (1983), "Premiums for High Quality Products as Returns to Reputations", *Quarterly Journal of Economics*, XCVIII, pp. 659-79.

SILVERGLADE, B. (1998), "Should the SPS Agreement be Amended? A Modest Proposal to Restore Public Support", Ceres Conférence on Politicizing Science: What Price Public Policy ? Georgetown University Public Policy Institute, 4 April.

SMITH, H. and R. LATTIMORE (1997), "The Search for Rules for Non-Tariff Barriers: Fire Blight of Apples", International Agricultural Trade Research Consortium meeting, San Diego, 14-16 December.

SPENCE, M. (1975), "Monopoly, Quality and Regulation", *Bell Journal of Economics*, 6, 2, pp. 417-29.

STIGLER, G. (1975), *The Citizen and the Stage,* Chicago University Press.

SYLVANDER, B. (1996), "Normalisation et concurrence internationale : la politique de la qualité alimentaire en Europe", *Économie rurale*, 231, pp. 56-61.

THILMANY, D.D. and C.B. BARRETT (1997), "Regulatory Barriers in an Integrating World Food Market", *Review of Agricultural Economics*, 19, 1: 91-107.

THORNSBURY, S., D. ROBERTS, K. DEREMER and D. ORDEN (1997), "A First Step in Understanding Technical Barriers to Agricultural Trade", The XIII International Conference of Agricultural Economists, August 10-16, Sacramento, California.

TULLOCK, G. (1997), "The Political Economy of Administered Decisions: What we Might Hope for and What we Can Expect", *Understanding Technical Barriers to Agricultural Trade*, D. Orden and D. Roberts, eds, The International Agricultural Trade Research Consortium, St Paul, Minnesota.

UNITED STATES INTERNATIONAL TRADE COMMISSION, *Fiscal Year 1997 Annual Report*, USITC publication 3085, Washington, D.C.

VISCUSI, W.K. (1993), "The Value of Risks to Life and Health", *Journal of Economic Literature*, vol. XXXI, December, pp 1912-1946.

VISCUSI, W.K. (1997), "Alarmist Decisions with Divergent Risk Information", *The Economic Journal*, 107, November, pp 1657-70.

VISCUSI, W.K., J.M. VERNON and JR J.E. HARRINGTON (1995), *Economics of Regulation and Antitrust,* The MIT Press, Cambridge, Mass.

VOGEL, D. (1995), *Trading up: Consumer and Environmental Regulation in a Global Economy,* Harvard University Press, Cambridge, Massachusetts.

OECD PUBLICATIONS, 2, rue André-Pascal, 75775 PARIS CEDEX 16
PRINTED IN FRANCE
(51 1999 08 1 P) ISBN 92-64-17159-2 – No. 50965 1999